# Contentious Data in Movement

This book explores the profound transformations brought about by the datafication of society, and reflects on the implications this has for activism, social movements, and contentious politics.

The result is a collection of chapters that advance the field of social movement studies theoretically and empirically, enabling us to better understand these transformations and offering a vocabulary and conceptual apparatus that facilitates a truly interdisciplinary dialogue. Through rich case studies, empirical examples, novel insights, and provocative reflections, the book serves as an invitation for scholars and activists to reflect on the theoretical, empirical, methodological and ethical implications of the datafied society, and its consequences for social movement activism.

The volume will be of great interest to scholars and researchers of social movements, political science, social anthropology, and ethnography. The chapters in this book were originally published as a special issue of *Social Movement Studies*.

**Cristina Flesher Fominaya** is Professor of Global Studies, Aarhus University, Denmark. She is Editor-in-Chief of *Social Movement Studies* and founding editor of *Interface Journal*. Her latest books are *Democracy Reloaded* (2020) and *Social Movements in a Globalized World* (2020).

**Stefania Milan** works at the intersection of political participation, technology, and governance, with emphasis on infrastructure and agency. She is Professor of Critical Data Studies at the University of Amsterdam, Netherlands, affiliated with the Berkman Klein Center for Internet & Society (Harvard University, Massachusetts) and the School of Transnational Governance (European University Institute, Italy). She was the Principal Investigator of the DATACTIVE project (2015–21), funded by a Starting Grant of the European Research Council.

**Davide Beraldo** is Assistant Professor at the Institute for Logic, Language and Computation and at the Department of Media Studies, University of Amsterdam, Netherlands. His research lies at the intersection of new media studies, computational social science, and the epistemology of complexity.

# Contentious Data in Movement

*Edited by*
**Cristina Flesher Fominaya, Stefania Milan and Davide Beraldo**

Routledge
Taylor & Francis Group
LONDON AND NEW YORK

First published 2025
by Routledge
4 Park Square, Milton Park, Abingdon, Oxon OX14 4RN

and by Routledge
605 Third Avenue, New York, NY 10158

*Routledge is an imprint of the Taylor & Francis Group, an informa business*



*British Library Cataloguing in Publication Data*
A catalogue record for this book is available from the British Library

ISBN13: 978-1-032-89055-5 (hbk)
ISBN13: 978-1-032-89056-2 (pbk)
ISBN13: 978-1-003-54102-8 (ebk)

DOI: 10.4324/9781003541028

Typeset in Minion Pro
by Newgen Publishing UK

**Publisher's Note**
The publisher accepts responsibility for any inconsistencies that may have arisen during the conversion of this book from journal articles to book chapters, namely the inclusion of journal terminology.

# Contents

# Citation Information

The following chapters in this book were originally published in *Social Movement Studies*, volume 23, issue 3 (2024). When citing this material, please use the original page numbering for each article, as follows:


**Chapter 1**

*Data in movement: the social movement society in the age of datafication*
Stefania Milan and Davide Beraldo
*Social Movement Studies*, volume 23, issue 3 (2024), pp. 265–284

**Chapter 2**

*"The future of the internet hangs in the balance": the perception and framing of political opportunity and threat in the contentious politics of data*
Jared M. Wright
*Social Movement Studies*, volume 23, issue 3 (2024), pp. 285–302

**Chapter 3**

*Amplification, evasion, hijacking: algorithms as repertoire for social movements and the struggle for visibility*
Emiliano Treré and Tiziano Bonini
*Social Movement Studies*, volume 23, issue 3 (2024), pp. 303–319

**Chapter 4**

*When digital capitalism takes (on) the neighbourhood: data activism meets place-based collective action*
Vassilis Charitsis and Mikko Laamanen
*Social Movement Studies*, volume 23, issue 3 (2024), pp. 320–337

**Chapter 5**

*Doubt to be certain: epistemological ambiguity of data in the case of grassroots mapping of traffic accidents in Russia*
Dmitry Muravyov
*Social Movement Studies*, volume 23, issue 3 (2024), pp. 338–354

**Chapter 6**

*Coordinating and doxing data: Hong Kong protesters' and government supporters' data strategies in the age of datafication*
Yao-Tai Li and Katherine Whitworth
*Social Movement Studies*, volume 23, issue 3 (2024), pp. 355–372

**Chapter 7**

*Datafication and implicated networks of demobilization: social movement demobilization in datafied societies*
Chi Kwok and Ngai Keung Chan
*Social Movement Studies*, volume 23, issue 3 (2024), pp. 373–390

**Chapter 8**

*Achieving Organizationality Through Authorship Affordances — A Communicative Episode of Telegram Polling from 2019 Hong Kong*
Marilyn Poon
*Social Movement Studies*, volume 23, issue 3 (2024), pp. 391–412

**Chapter 9**

*PROFILE: Revisiting* the social movement society *in a time of datafication*
David S. Meyer
*Social Movement Studies*, volume 23, issue 3 (2024), pp. 413–421

**Chapter 10**

*Data as narrative: contesting the right to the word*
Nick Couldry
*Social Movement Studies*, volume 23, issue 3 (2024), pp. 422–428

# Notes on Contributors

**Davide Beraldo** is Assistant Professor at the Institute for Logic, Language and Computation and at the Department of Media Studies, University of Amsterdam. His research lies at the intersection of new media studies, computational social science, and the epistemology of complexity.

**Tiziano Bonini** (PhD, University of Siena) is Associate Professor in Sociology of Communication and Culture at the Department of Social, Political and Cognitive Sciences at the University of Siena, Italy. His research interests include political economy of the media, platform studies, critical algorithm studies, media production studies, and digital cultures

**Ngai Keung Chan** is Assistant Professor at the School of Journalism and Communication at The Chinese University of Hong Kong. His research examines the intersection of platform governance, algorithms, and service work. His work has appeared in such journals as *New Media & Society*, *Information, Communication & Society*, *Surveillance & Society*, *Space and Culture, and Media and Communication*, among others.

**Vassilis Charitsis** is a lecturer at Brunel Business School, United Kingdom. His research interests lie in the intersection of digital capitalism and the datafication of everyday life. His work has been published in *Television & New Media*, *Surveillance & Society*, *Triple C: Communication, Capitalism and Critique*, *Marketing Theory*, and *Ephemera: Theory and Politics in Organisation*.

**Nick Couldry** is a sociologist of media and culture. He is Professor of Media Communications and Social Theory at the London School of Economics and Political Science, and, since 2017, also a Faculty Associate at Harvard's Berkman Klein Center for Internet and Society, Massachusetts.

**Cristina Flesher Fominaya** is Professor of Global Studies, Aarhus University, Denmark. She is Editor-in-Chief of *Social Movement Studies* and founding editor of *Interface Journal*. Her latest books are *Democracy Reloaded* (2020) and *Social Movements in a Globalized World* (2020).

**Chi Kwok** is Assistant Professor in the Department of Government and International Affairs at Lingnan University in Hong Kong. His current research interests include political economy, corporate power, the gig economy, and workplace justice. His work has appeared in journals such as, among others, *Journal of Social Philosophy*, *Philosophy & Social Criticism*, *Review of Social Economy*, and *Information, Communication & Society*.

**Mikko Laamanen** is Associate Professor at the LifeStyle Research Center of Emlyon Business School, France. His research focusses on the everyday politics of inclusion and social change, with empirical work on consumer lifestyle movements, organizing alternative economies, and outreach in arts. His research has been published in, amongst others, *Current Sociology*, *International Journal of Consumer Studies*, *Journal of Cleaner Production*, *Marketing Theory*, and the *Springer Handbook of Social Movements Across Disciplines*.

**Yao-Tai Li** is Lecturer of Sociology and Social Policy in the School of Social Sciences at University of New South Wales, Australia. He holds a PhD in sociology from the University of California, San Diego. His research interests include contentious politics, identity, and social media. His work has been published in several scholarly journals including *British Journal of Sociology*, *The China Quarterly*, *World Development*, *Urban Studies*, *New Media and Society*, *Big Data & Society*, *Work, Employment and Society*, *The Sociological Review*, *Ethnic and Racial Studies*, *Discourse & Society*, *Social Science Computer Review*, *Journal of Contemporary Asia*, among others.

**David S. Meyer** is Professor of Sociology and Political Science at the University of California, Irvine. He writes mostly about social movements, social change, and public policy. His most recent book is *How Social Movements (Sometimes) Matter* (2021).

**Stefania Milan** works at the intersection of participation, digital technology, and governance, with an emphasis on infrastructure and citizen agency. She is a Professor of Critical Data Studies at the University of Amsterdam, Netherlands, affiliated with the Berkman Klein Center for Internet & Society (Harvard University, Massachusetts) and the School of Transnational Governance (European University Institute, Italy). She was the Principal Investigator of the DATACTIVE project (2015–21), funded by a Starting Grant of the European Research Council.

**Dmitry Muravyov** is a social researcher who focuses on the intersections of the internet, data, knowledge, and politics. He is a PhD candidate at the Delft University of Technology, Netherlands, and a member of the grassroots research community Club for Internet and Society Enthusiasts.

**Marilyn Poon** is an organization studies scholar whose research focuses on the digital condition and its influence on organizing phenomena, including leadership and entrepreneurship. She applied a communication as constitutive or organizing (CCO) approach in her dissertation, which focused on the 2019 Hong Kong Protests to explore the role of platform technologies. Currently, she is applying practice-based theorizing to unpack the generative dynamics of data-driven and value-based practices.

**Emiliano Treré** (PhD, University of Udine) is Reader in Data Agency and Media Ecologies at Cardiff University's School of Journalism, Media and Culture. He is a widely cited author in digital activism and critical data studies with a focus on the Global South. He co-founded the 'Big Data from the South' Initiative and co-directs the Data Justice Lab. His book *Hybrid Media Activism* (2019) won the Outstanding Book Award of the ICA Interest Group 'Activism, Communication and Social Justice'. His latest book, *Algorithms of Resistance: The Everyday Fight against Platform Power*, with Tiziano Bonini (2023), explores power, agency, and resistance in the platform society.

**Katherine Whitworth** received her PhD in political economy from the University of Sydney, Australia. Her research interests include political legitimacy, social welfare, contentious politics, Hong Kong and Chinese politics. Her work has been published in scholarly journals such as *The China Quarterly*, *Big Data & Society*, *Urban Studies*, *World Development*, *Journal of Contemporary Asia*, *Journal of Consumer Culture*, *Social Science Computer Review*, and *International Migration*. She is currently conducting a project on the Lennon Walls with Yao-Tai Li.

**Jared M. Wright** is Assistant Professor at TED University in Ankara, Turkey, and Director of the Center for Digital Social Science (CDSS) virtual research network. He earned his PhD in sociology from Purdue University, Indiana, in 2020. His areas of research include digital sociology, social movements, and new media

# Introduction: *Contentious Data in Movement*

Cristina Flesher Fominaya

This volume brings together a set of articles that take as their starting point the profound social transformations brought about by the increased datafication of society, and reflect on the implications this has for activism and social movements. The result is a collection of articles that advance the field theoretically and empirically, enabling us to better understand these transformations and providing us with a vocabulary and conceptual apparatus that enables the field to move forward.

Rapid technological advancements have always posed a challenge for theory: how robust are consecrated theoretical frameworks in light of rapidly changing empirical realities? At what point do these transformations necessitate new concepts to understand them? How can we resist the temptations to reinvent the wheel, throw the baby out with the bathwater, or hop on the latest semantic bandwagon? Finding the genuinely new requires both critical imagination and restraint. At the simplest level we can "add on" categories to the forms that activism takes: we can chart forms of data activism, data justice, and resistance against the ways that AI-related processes and datafication enable increased mass surveillance and the development of algorithm-facilitated regimes of repression and oppression. This in itself is a much needed and important contribution to knowledge. Even more challenging is engaging our critical imagination to draw insights from a range of disciplines to make sense of how datafication not only becomes an object of resistance but traverses the very forms and imaginaries of movement activism and shapes the terrain on which activism manifests. The chapters in this volume, drawn from a special issue in *Social Movement Studies* conceived by Stefania Milan and Davide Beraldo, does both, drawing on empirically grounded case studies to reflect on activism and social movements in our current shifting datafied societies, while developing new concepts to help us make sense of these transformations. It serves as an invitation to scholars working in media studies, digital sociology and anthropology, critical data studies, surveillance studies, security studies, human–computer interaction studies, and AI-related art and scholarship who are also interested in navigating the media–technology–social movement nexus to join in this collective reflection and engagement.

Milan and Beraldo kick off the collection by offering a provocation on whether current social movement scholarship is fit for purpose in an increasingly datafied society. They do so by taking five social movement-related concepts and processes – group formation, opportunity structures, action repertoires, meaning work, and collective identity – and ask in what ways they "are altered by datafication and the advance of intelligent systems in society". Using myriad examples, they show how relevant movement interlocutors and the very spaces for participation are shifting and highlight the contradictory implications of these changes for movements. As public–private partnerships replace governments and become the primary holders of information on citizens, they erode established mechanisms of democratic accountability and citizen agency, disrupting established practices of resistance. Social media platforms, for their part, both enable and constrain activism in ways that are often invisible and unintelligible to the public (and to many activists as well). The wider implication of this comprehensive argument is that datafication, unlike digitalization, has a much more profound effect on society because it has the capacity to *direct* social behavior and interactions (e.g., the processes through which algorithms can direct behavior such as consumption choices or opinion formation). As such, datafication does not affect only movements specifically concerned with "data activism" but all social movements and forms of activism.

In another theoretically rich contribution, Wright delves deep into social movement scholarship by exploring how data affect how political opportunities and threats are perceived and mediated by activists engaged in contentious data politics. Wright argues on the basis of two case studies (the Digital Rights Movement and Anonymous) that while traditional social movement scholarship emphasizes the importance of perceived political opportunity as a motivator for action, threat frames were more prominent in the context of the "Internet freedom" struggles, with opportunity frames having less salience for online activism. Wright stresses this finding was not only due to the structural form of online activism but also to cultural factors – an important reminder that online activist practices (like all activism) is embedded in specific cultural milieus.

Treré and Bonini also contribute a theoretical framework that helps us make sense of these changes. They articulate algorithmic activism as a subset of algorithmic politics, and offer a typology of three dynamics of algorithmic activism: *amplification*, *evasion*, and *hijacking*. *Amplification* involves hacking algorithms to amplify movement narratives. Activists use *evasion* to try to bypass or avoid algorthmic moderation – a strategy often used by the far right and other groups subject to censorship or banning of hateful or harmful messaging (e.g. pro-anorexia messages). *Hijacking* – often but not exclusively applied to hashtags – can be used for both social justice and repressive purposes by institutional and movement actors as they engage in struggles over the control and shaping of public space on and offline. This rich and clear contribution should be of great value to anyone seeking to categorize and analyze different strategies and tactics of algorithmic activism and politics.

Charitsis and Laamanen address data activism against data colonialism by Big Tech (a process known as *Googlization*), focusing on two case studies: the creation of a Google campus in Berlin and the Sidewalk Toronto project (also led by Google). These cases brought together neighborhood activism against gentrification and housing (place-based activism), with activism around Big Tech's unethical data practices, putting forth alternative visions of "human-centric technologies and data politics". The case studies provide novel insights into the interaction between online and offline politics, and show how data activists are working to highlight how datafication is not just composed of unintelligible and opaque processes but are "manifested in everyday life in very concrete ways". Charitsis and Laamanen appeal to social movement scholars studying ICT use to move beyond their material and instrumental aspects to consider their symbolic importance, with an emphasis on data (and digital capitalism more broadly) as a contentious issue connected with other processes of social inequality and the disempowerment of citizens and communities.

In his contribution, Muravyov addresses the relation between civic technology initiatives and data activism in the Russian context. He specifically analyzes data activist practices around a geoweb mapping project with the goal of reducing traffic accidents. Data activists engage with official government data, adopting an "epistemological ambiguity" in their grassroots mapping activity. At times the data is used as the basis for proposing changes, at others the data is questioned and corrected drawing on interactive citizen engagement. Simultaneously using, questioning, and improving data develops and sustains users' trust while users themselves actively contribute to the quality of the project by supplying missing data and correcting flaws in the official data, offering an alternative vision and practice of how road accidents can be reduced.

Li and Whitworth juxtapose protester and government supporter strategies in the age of datafication in the case of Hong Kong's Anti-ELAB movement, demonstrating how data can form parts of movement repertoires as well as becoming a contentious issue in its own right. In the context of Hong Kong, in which access to data is framed as a service or a privilege rather than as a right by the government, data activists have begun building their own datasets and contesting the norms around the openness of data as well as offering new imaginaries around data-driven governance and data practices. Focusing specifically on the practice of doxing, they argue that this practice sits at the nexus of Beraldo and Milan's (2019)[1] dichotomy between "data as weapon" and "data as stakes". They argue that the overlap in this dichotomy is particularly relevant in this movement because it appears in the contentious repertoires of activists on both sides of the struggle around the anti-extradition law amendment bill (Anti-ELAB) of 2019. The collection of protester data through facial recognition software, transport usage records, and social media posts formed a key part of the "soft strategies" used by the Hong Kong government to repress and police protesters. In response, protesters resisted by wearing masks, destroying CCTV cameras, and disabling facial recognition logins on their telephones, thereby disrupting the collection of biometric information by the government. In addition, they weaponized data through the practice of doxing, i.e., publicly exposing private information without consent), targeting police and people suspected of being informants. In response, police also began wearing masks and stopped displaying their badge numbers in order to conceal their identity from the protesters, while government supporters began to collect and publicize images of individuals participating in unsanctioned protests and other illegal activities. This fascinating case study provides an in-depth reflection on the complex reciprocal practices of surveillance, anti-surveillance and sousveillance, as well as provoking reflexivity around our understanding of what ethical data practices are and under what circumstances practices like doxing are justifiable.

Kwok and Chan's contribution engages with a crucial yet understudied aspect of social movement scholarship – demobilization – in order to introduce the theoretical framework of *implicated networks of demobilization* as a means of understanding this process in datafied societies. Demobilization processes reflect a reduction in the aggregation of resources available to political actors who are seeking to make claims collectively (Tilly and Tarrow 2015)[2]. States can take advantage of data infrastructures to demobilize protest movements through such practices as doxing or creating information gluts. Drawing on examples from the Anti-ELAB movement in Hong Kong and the 2020 George Floyd protests in the United States, their framework attempts to understand data-mediated dynamics between the state and activists. They offer three key dimensions of their framework: *implicated data dynamics* that recognizes that protest coordination and communication are increasingly conducted online and leave relatively durable and traceable imprints for policing that enables the state to target and monitor activists for reactive and preventive demobilization; *materiality dynamics* that are centered around struggles over material objects and infrastructures; and *epistemic dynamics* that highlight the importance of mastery of digital knowledge and tools to protect against potential arrest, particularly with regard to state surveillance. In addition to providing a rich set of examples from these two movements to illustrate the theoretical framework, the authors show how their framework could be a useful tool for further research on digitally mediated struggles and the dynamic interplay between mobilization and demobilization in social movements.

Also drawing empirically on the Anti-ELAB movement in Hong Kong, Poon's contribution delves into activists' "backstage" organizing practices through an exploration of their use of the affordances of Telegram's polling feature to take collective decisions. Poon shows how Telegram's polling feature facilitated not only collective decision-making but also cultivated the movements' collective identity. Although the polling practices enabled moving from grievances to action, the affordances of the app prevented activists from tangibly understanding the consequences of their actions and limited their ability to see the fuller picture when voting on the app. Poon's exploration of how Telegram's chat app influenced data-related practices in the movement unearths some of the more troubling aspects of data-mediated activism, while contributing novel insights on the processual dynamics of authority practices.

In light of these articles and the issues brought about by datafication, Meyer revisits the assumptions and implications of the concept of the "social movement society" that he and Tarrow put forth in 1998.[3] This invited contribution is a good example of how empirical changes in lived reality push us to reconsider what we need to keep and what we need to modify in our conceptual and theoretical armories. As Meyer argues, he and Tarrow could not have foreseen the ways changes in communication technologies would foster the emergence and growth of populist movements. Meyer draws on examples from Occupy Wall Street to show how these communication technology advances reduce the need for certain forms of expertise and resources, meaning that activism can in a sense precede resources and bypass institutional information filters, fostering and expanding movements' dissemination potential. Meyer's discussion highlights how datafication has changed the stakes and targets of mobilization, and has created new grievances, tools, and coalitions of activists.

In the final invited contribution, Couldry reflects on the fundamental processes underlying data activism, drawing on the collected articles for inspiration. He points to three key changes: a change in the general conditions under which all social movements operate; data itself becoming either the specific or general object of activism; and data becoming crucial to movement practices of resistance. He then draws on Melucci to argue that "data as narrative" is becoming an increasingly important aspect of contemporary politics. Couldry uses Melucci's (1996)[4] assertion that "real domination is today the exclusion from the power of naming" to reflect on the ways that datafication has led to the increasing centrality of struggles over the "right to the word", a struggle in which data is encoded with particular interests and power relations that are then contested by resistance from below. He closes by arguing that focusing on the specifics of instrumental and technological affordances should never lead us to lose sight of what is truly at stake in processes of contention and resistance, echoing a reflection in critical social media/movement scholarship that is also reflected in many of the contributions here.

Taken together, the contributors in this volume offer a compelling invitation to social movement scholars and activists to meet datafication head on in all its promise and ugliness, its power and its threats, and offer some powerful tools and insights to face that challenge. It has been my pleasure and privilege to play a part in this project, and I have learned a tremendous amount from all of these contributions, which have changed the way I view contemporary society and contentious politics. I hope it may do the same for you.

Aarhus, May 2024

## Notes

1. Beraldo, Davide, and Stefania Milan. "From data politics to the contentious politics of data." *Big Data & Society* 6.2 (2019): 2053951719885967.
2. Tilly, Charles, and Sidney G. Tarrow. *Contentious Politics*. Oxford University Press, 2015.
3. Meyer, David S. and Sidney Tarrow (Eds.). *The Social Movement Society: Contentious Politics for a New Century.* Rowman and Littlefield, 1998.
4. Melucci, Alberto. *Challenging Codes*. Cambridge University Press, 1996.

# Data in movement: the social movement society in the age of datafication

Stefania Milan and Davide Beraldo

**ABSTRACT**

Data has increased currency in contemporary activism and has become an integral part of the action repertoire of today's social movements. But data and data infrastructure are not only tools to support a movement's struggle: they are constitutive parts of the environment in which movements operate, and objects of contention in their own right. These developments challenge scholars of social movements and collective action. How are movements changing under the pressure of datafication? What 'new' mechanisms, actors and tactics meet the growing demand for citizen participation in an increasingly datafied society? Is social movement scholarship 'fit' to capture and interpret this evolution? This theoretical article puts social movement studies in dialogue with critical data studies with the aim of encouraging a much-needed cross-pollination. It advances the notion of 'datafied movements' to address the novel structural condition of contentious politics in the age of datafication and to explore the socio-technical, systemic effects of data and data infrastructure on movement dynamics. It reflects on how five key social movement dynamics, and the related elements in the conceptual toolbox of social movement studies – group formation, opportunity structures, action repertoires, meaning work, and collective identity – are altered by datafication and the advance of intelligent systems in society. In so doing, this article charts the building blocks of a future-proof research program in social movements.

## Introduction

Using do-it-yourself open-source sensors, citizens of the Colombian capital Bogota have been collecting their own evidence on air pollution since 2017. They 'leveraged citizen data to participate in the public debate on air pollution' and 'challenged government data infrastructures through prototyping alternatives' (Barreneche & Lombana-Bermudez, 2023, p. 3639). In China, a group of activists crowdsourced an archive of sexual violence, using data-scraping tools to organize stories uploaded by ordinary citizens into a database in view of 'pressuring the government to reform outdated laws' (Sun & Yin, 2022, p. 8). Data contribute to 'fighting bias, building progressive movements, and promoting civic engagement,' claims Data for Black Lives, a branch of the Black Lives Matter movement mobilizing around the slogan 'Data as protest. Data as accountability. Data as collective action' (Data for Black Lives, n.d.). These initiatives are amongst the many that 'rely on objectivity as the lynchpin to provide truthful accounts to decision makers and contest rhetoric and practices that oppress' (Meng & DiSalvo, 2018, p. 7). They expose how data has gained increasing currency in contemporary activism.

But social movements are also ever more interested in the infrastructure that enables the creation and analysis of data, henceforth data infrastructure. During the 2019 Hong Kong pro-democracy protests, activists were seen taking down 'smart lampposts' feared to deploy facial recognition technology (Wong, 2020). When the then US president Donald Trump threatened with deletion federal evidence on climate change, a coalition of scientists and volunteers embarked in an act of 'data resistance' self-archiving scientific information in danger of erasure (Vera et al., 2018). Not only are data and data infrastructure tools supporting movement narratives: they have become objects of contention in their own right.

We live in a datafied society, characterized by the transformation of many aspects of human and social life into machine-readable data – a process known as 'datafication'. Think of the quantification of emotions and friendships afforded by social media platforms, or the evidence-based policymaking touted by public administrations within 'smart city' projects. Datafication represents the last evolution in the 'computational turn' initiated in the 1950s. Computation has massively accelerated over the last two decades, with unprecedented volumes of information ('big data') being generated and processed to produce value. This shift is 'not merely technological, but also social and political, and it therefore confronts us with questions of power, agency and control' (Hintz et al., 2018, p. 2). Datafication is now firmly installed 'at the core of our culture and social organization' (van Es & Schäfer, 2017, p. 14). It alters the dynamics of political participation and mobilization, affecting social movements in significant ways, as the examples introducing this article illustrate.

These developments at the crossroads of the technological and the social dimensions of political participation pose several questions for scholars of social movements and collective action. How are movements changing under the pressure of datafication? What sort of 'new' mechanisms, tactics, actors meet the growing demand for citizen participation in an increasingly datafied society? Is social movement scholarship 'fit' to capture and interpret this evolution?

This article begins with the observation that if datafication changes the way in which people get together, what they can do and how they do it, it affects the very same foundations of organized collective action. Consequently, the discipline of social movement studies is at a crucial turning point. For one, datafication has multiplied the occasions for social actors to engage in the promotion of social change. For instance, 'open data' made available by public administrations allow social justice activists to scrutinize state output and direct their campaigning efforts (Torres, 2019). These developments facilitate the emergence of new coalitions, innovative campaigns, and original action repertoires (Gutiérrez & Milan, 2019). They contribute to making social movement studies relevant to novel audiences – including scholars of media, culture, political communication, political systems, and governance – promoting inter-disciplinary collaborations. On the other hand, social movement studies face a grand challenge: if the discipline is to successfully navigate the 'paradigm shift' brought about by datafication (Kitchin, 2014), scholars can no longer ignore data-related dynamics, and their material (e.g., data-production mechanisms) and symbolic dimensions (e.g., meanings attributed to data by social actors). This might require updating and expanding the vocabulary of the discipline, making room for 'better integration of the different intellectual traditions that are currently focused on the same set of empirical phenomena' (Flesher Fominaya & Gillan, 2017, p. 383). This is precisely the goal of this article and the special issue of which it forms a part.

Exploring these claims, we put social movement studies in dialogue with critical data studies. Critical data studies is an interdisciplinary field of inquiry concerned with data as (power) relations. It investigates the premises and effects of data practices and infrastructures, rather than naturalizing them as a matter of empirical inquiry or as a methodological device (Iliadis & Russo, 2016). In promoting this exchange, we invite a much-needed cross-pollination of vocabulary and research programs.

This conceptual article is organized as follows. First, we explain what is at stake with datafication and how this represents a paradigm shift that goes beyond digitalization. Second, we argue that these developments affect social movements as a whole, not just the specific phenomenon of data activism (that is, activism taking a critical approach to data). We introduce the notion of 'datafied movements' to capture the novel conditions under which organized collective action unfolds today. Finally, we put this notion to use to identify five movement dynamics that are altered by datafication: group formation, opportunity structures, action repertoires, meaning work, and collective identity. We conclude by encouraging social movement scholars to pay closer attention to the effects of the materiality of data on mobilization dynamics.

## From digitalization to datafication

Datafication refers to the 'quantification of human life through digital information, very often for economic value' (Mejias & Couldry, 2019). Albeit sometimes conflated, datafication differs from digitalization, or the transformation of organizational processes in virtue of the adoption of digital technologies, in that it is a 'deeper' intervention in the material and social foundations of reality. Datafication is an active intervention in the material and social foundations of reality-one *able to steer* such reality. The impact of the logic of social media platforms on users' behavior is a case in point (van Dijck & Poell, 2013). What's more, datafication is a pervasive process that leaves little room for disengagement and affects individual life and social relations far beyond media use. By way of example, think of platforms like TikTok or Instagram. Not only do they enable new forms of self-expression and media consumption; they also generate, accumulate and circulate data about the lived experiences of users, making room for new forms of value extraction, intelligence, and social engineering. A handful of tech firms like Google/Alphabet and Facebook/Meta constitute the engine of this socio-economic system based on the systematic exploitation of personal data (Srnicek, 2016).

Datafication also increasingly permeates state functions and mediates state–citizen relations – with social consequences. One example is predictive policing enabled by algorithms, which targets citizens across a range of previously disconnected datasets and institutions, (re-)producing patterns of racial discrimination (Ferguson, 2017). Another is the deployment of machine learning in evidence-based policymaking, which leaves people with fewer mechanisms for appeal and redress (López, 2020). Social inequality is amplified as mechanisms of social protection deploy statistical models and predictive algorithms that are often imperfect (Eubanks, 2018), rendering invisible many communities at the margins (Milan et al., 2021). As a result, democratic norms – including the autonomy of representative institutions and the safeguarding of individual privacy in public space – are challenged by the 'distributed and largely uncontested new expression of power' (Zuboff, 2015, p. 75) enshrined in the global architecture of data commodification. Public–private partnerships replace governments as the primary holder of the monopoly over information on the citizenry, eroding established mechanisms of democratic accountability and ultimately citizen agency.

Datafication has also changed what constitutes the public sphere and participation today. For instance, social media platforms both empower and jeopardize people's ability to mobilize. They deploy recommender systems that cluster users based on their affinity, tailoring content to algorithmic predictions and increasing group homophily. But YouTube has been found to amplify extremist context online (Whittaker et al., 2021). And biometric facial recognition has a chilling effect on the right to peaceful assembly and protest (Zalnieriute, 2021) and resistance to it is on the rise (Madison & Klang, 2019).

Until recently, these developments have been invisible and unintelligible to the public. However, events like the 2013–2016 Snowden leaks (Hintz & Brown, 2017) and the 2018 Facebook-Cambridge Analytica scandal (Heawood, 2018) have contributed to bringing them under public scrutiny, creating awareness about the centrality of data and data infrastructure in our life. These incidents have laid bare the extent to which 'data politics' is rife with socio-political consequences (Ruppert et al., 2017).

In outlining an analytical approach to study collective action in the datafied society, we take the lead from the foundational article 'Navigating the technology-media-movement complex' by Flesher Fominaya and Gillan (2017). Attempting a similar 'field-definition' operation, we observe that movement engagement with datafication does not emerge in a vacuum. But contrary to the authors' understanding of digitalization as an evolution in continuity with the past with respect to activists' self-expression, network formation, etc., we argue that datafication as a pervasive social order breaks with the past. We contend that datafication represents more than a mere change in the media and communication channels through which movements speak (cf. Cammaerts et al., 2013). Data-mediated relations do not only entail what is commonly understood as (intentional) communication processes, as they are also generated as the subproduct of everyday actions (e.g., carrying around a geolocated smartphone) or administrative acts (e.g., being registered in a healthcare database). We thus encourage movement scholars to move beyond the understanding of infrastructure development as 'mere background' (Flesher Fominaya & Gillan, 2017, p. 386). This is because data introduce novel (power) relations between the lived experiences of people, including movement actors, and the complex socio-technical assemblages that mediate them. Derived from Science and Technology Studies, the notion of socio-technical assemblages (cf. Jasanoff, 2017) is also central to critical data studies. It stresses the entanglement between the social and the technological dimension of an artifact or social phenomenon. Think of movements using social media to mobilize supporters: the algorithmic environment of the platform influences who receives the message and in relation to what other content it is served. In this equation, movements are empowered to reach out to novel audiences, but they are also stripped of control over the circulation of their messages. Moreover, data are not self-standing technological objects: they do not have a proper ontological status outside the processes, practices and technologies defining and distributing them. In other words, there are no data without devices, dashboards, algorithms, or data analytics; data are embedded in a complex assemblage (Mattoni & Pavan, 2018). Data invite us to adopt an approach to the study of contentious politics that is simultaneously systemic (i.e., not reducible to individual conditions) and socio-technical (i.e., waving together the social and the technological dimensions). From this we derive two observations.

First, today, data and related technologies are the inescapable backdrop against which social dynamics emerge and operate. Datafication is a 'totalizing' structural process (Couldry, 2020), which permeates social activity, including political contention, on a global scale and in more and more domains. Intervening also beyond people's intentionality and/or awareness, data contribute to set the stage for opportunities and motivations to act. Not only do they inform activists' tactics; they also provide new reasons to mobilize and reconfigure the space in which movements operate.[1] This is why in our study of social movements we ought to move beyond an instrumental relation to technology (cf. Treré, 2016), data and data infrastructure.

Second, data are at once material and cultural arrangements. On the one hand, data change the processes of aggregation and evolution of social formations at a very material level. The core mechanisms of datafication, e.g., algorithmic visibility, define novel 'infrastructures of the collective', that is to say 'institutional arrangements that enable the convergence of a collective in the first place, that also organize the circulation of information, ideas and people' (Dolata & Schrape, 2016, p. 5). This is exemplified by emerging data-mediated environmental activism in the Brazilian Amazon where environmental activists collaborate with 'info-mediators' to create meaning for the broader public (Rajão & Jarke, 2018). On the other hand, data and the associated dynamics are cloaked with an aura of objectivity and authority (or 'dataism', van Dijck, 2014) that affords them a distinctive rhetoric power in contemporary societies – which may influence movement framing and identity-making efforts.

In exploring these claims, we expose what is at stake for contemporary movements and their students.

## The advent of datafied movements

In 1998, Meyer and Tarrow hypothesized the emergence of a 'social movement society', where social protest had become a routinized element within conventional politics in advanced industrial democracies. Movements, they claimed, are an 'historical and not a universal way of mounting collective claims', apt to articulate dissent 'when the consolidated nation-state assured its citizens regular means of communication, created standard but fungible identities, and provided challengers with uniform targets and fulcra for acting collectively' (Meyer & Tarrow, 1998, pp. 4–5). But what happens to movements when the background in which they operate is so dramatically altered by datafication? To tackle this question, we advance the notion of 'datafied movements' to indicate the socio-historical configuration of the social movement society in the age of datafication. Contrary to notions like 'media activism', 'digital activism' or 'data activism', with datafied movements we do not aim to identify a particular subset of collective action. Our notion does not (only) relate to movements explicitly directed towards datafication as an issue of contention or incorporating datafied tactics in their action repertoire. Instead, it points to how virtually any type of social movement today evolves vis-à-vis the totalizing effects of datafication.

The state of the art we briefly review below reveals how social movement scholars have to date analysed separately the role of information and technology-two constitutive elements of 'big data'. Concerning the role of information, we know that its strategic mobilization is central to a movement's effort to frame its claims to stir collective action (Noakes & Johnston, 2005; cf.; Keck & Sikkink, 1998). Information allows movements to craft their media representation and their communication strategies (e.g., Mattoni & Treré, 2014; Neumayer & Rossi, 2016; Ryan & Jeffreys, 2019). The advent of social media represented a game changer (e.g., Barassi & Zamponi, 2020; Poell, 2020), and today movements favour a 'hybrid' engagement with the media landscape (Treré, 2019). They develop specific 'movement media cultures', defined as the set of tools, skills, practices, and norms developed by movement actors (Costanza-Chock, 2012). Furthermore, in relation to their use of information, movement actors have been approached as framing entrepreneurs (Snow, 2004), knowledge producers (Della Porta & Pavan, 2017) and bearers of 'counter-public knowledge' (Hess, 2016). From an infrastructural perspective, technologies of information production and dissemination have been of interest to scholars of media activism (e.g., Carroll & Hackett, 2006; Meikle, 2002) and digital activism (e.g., Milan & Barbosa, 2020; Pickard & Yang, 2017). The emergence from the late 1990s of global networks of counter-information like the Independent Media Center (Indymedia) contributed to the self-narrative of social movements and to the articulation of a novel 'cultural politics of technology' (Pickard, 2006, p. 315). Non-profit communication infrastructure like radio stations and servers created by tech-savvy activists served as the digital backbone of movements (e.g., Atton, 2004; Dunbar-Hester, 2014; Milan, 2013). Digitalization has resulted in the acceleration of protest cycles (Hardt, 2017), transforming mobilization and opportunity structures (Garrett, 2006; see also Kaun & Uldam, 2018). Scholars have exposed the 'institutional role' of technological infrastructures on social formations (Dolata & Schrape, 2016, p. 2; see also González-Bailón et al., 2011). They have noted how social media empower specific types of collective action, notably forms of 'personalized' public engagement such as 'connective action' (Bennett & Segerberg, 2012) and 'cloud protesting' (Milan, 2015a).

We contend that this literature, however rich, fails to grasp the dramatic transformation of public life promoted by datafication. First, social movement literature tends to adopt an instrumental approach to both information and technology, which privileges what movements do with it at the expenses of the sociotechnical mechanisms underlying their production and circulation. But this is increasingly untenable in the age of datafication. Second, information and technology taken together have received little attention. We claim that bracketing data as information (content) and data as technology (infrastructure) is a necessary premise, and the starting point, for a theory of datafied movements. We need to expand the boundaries of social movement studies if we are to grasp the relation between movements and their datafied environment, overcoming the dominant instrumental approach.

The notion of 'contentious politics of data', pointing to bottom-up initiatives that interfere with dominant processes of datafication (Beraldo & Milan, 2019), takes a step forward in this direction. It is visible in an array of data activism practices, where data are either constitutive of a novel action repertoire or mobilized as objects of struggle. Interpreting datafication as threat and/or opportunity, data activists seek to resist ('reactive data activism') or repurpose ('proactive data activism') data and data infrastructure (Milan & van der Velden, 2016). Examples include encryption activists training human rights defenders in protecting their communications (Kazansky, 2021), civic tech communities appropriating open data for social change (Wissenbach, 2020), or citizens re-elaborating official government data to change the official narrative around traffic accidents (see Muravyov, 2022, in this issue).

But while the heuristic of data activism can help us in understanding a certain type of engagement of movement actors with their datafied environment, it can only partially account for the broad transformations caused by datafication and impacting a movement's constitution, evolution, and effectiveness. The paradigm shift brought about by datafication has consequences so significant as to alter the dynamics and meaning of citizenship, political participation, and by extension of organized collective action. In other words, data and data infrastructure are constitutive of 'environments within which [people] move and that shape the structure of their perceptions, their forms of discourse, and their social behaviour patterns' (Heise, 2002, p. 151).

To stress the systemic and socio-technical implications of datafication on social movements, we advance the notion of 'datafied movements'. Datafied movements do not represent a particular class of activism (such as data activism), rather the latest 'historical' (Meyer & Tarrow, 1998, p. 4) evolution of social movements vis-à-vis the changing structural conditions of contentious politics in the age of mass surveillance and algorithmic mediation. To 'give legs' to this notion, we now proceed to reread five key social movement elements and dynamics – group formation, opportunity structures, action repertoires, framing and meaning work, and collective identity – that are challenged and changed by datafication, and that we see, separately or in combination, as archetypical of datafied movements. In making explicit the datafied foundations of today's movements, we start outlining a research program of recalibration of classical social movement concepts.

## Rethinking social movement studies for the datafied society

The recent history of the discipline has seen several attempts to reconceptualize collective action for a shifting (digital) media environment, retrofitting key concepts to future-proof them (see, among others, Bennett & Segerberg, 2012; Bimber et al., 2012; Kavada, 2016; Mortesen et al., 2019; Trottier & Fuchs, 2014). In this section, we add to this trend by reflecting on *the ways in which datafication intervenes in the structural, agential and symbolic dimensions of social movements.*

### *Algorithmic group making*

The first significant change brought forward by datafication concerns group formation, including recruitment and coalition building. We speak of algorithmic group making to indicate the impact of the processes, some of which are detailed below, that affect the way people coalesce in a movement-like formation in the datafied society. Specifically, we observe three trends.

#### *Platform-mediated recruitment and leadership*

Non-human, datafied mediators like digital objects (e.g., hashtags) and procedures (e.g., algorithms) re-shape movement recruitment and leadership dynamics. The mechanisms of interpellation built in in social media (e.g., tags, mentions, recommender systems suggesting 'friends' or content one may like) allow activists to directly appeal to publics that were once harder to reach (Milan, 2015b). Known leadership patterns are complemented by the 'choreography of assembly' enabled by platforms, whereby 'influential Facebook admins [. . .] become "soft leaders" or choreographers, involved in setting the scene, and constructing an emotional space within which collective action can unfold' (Gerbaudo, 2012, p. 5; see also Poell et al., 2016). Moreover, non-human actors (cf. Latour, 2005) like hashtags play a novel role as brokers in movement networks (Beraldo, 2022b), calling for an update of human-centric network approaches to movement formation (see Diani & Mische, 2015).

#### *Groups as algorithmic outcomes*

As a result of the algorithmic mediation of the activist agenda, movements may approach algorithmic outcomes governed by the rules of visibility and aggregation dictated by corporate platforms (Milan, 2015b) – and their strategic re-appropriation by movements themselves. They can be seen as 'calculated publics' (Gillespie, 2014), or issue groups that are brought together by the automated procedures of platforms. These novel logics of aggregation are not merely the consequence of the empowering effect of the 'costless' technology of communication and organization, as the notions of 'networked publics' (Papacharissi, 2010), 'mass self-communication' (Castells, 2009), or 'connective action' (Bennett & Segerberg, 2012) seem to suggest. Rather, platform metrics (Beer, 2016), such as the real-time measurement of content popularity which determines the likelihood an item is displayed to users and its prominence on the users' interface, play a crucial role in this process.

#### *Accelerated mobilization and demobilization*

Mobilization (and demobilization) dynamics are altered by the modified logic of visibility of an environment characterized by real-time metrics, chains of sharing, and algorithms rewarding virality. This can result in a scenario of the type 'faster is different' (Tufekci, 2011): the formation of a (visible) critical mass of participants, a precondition for successful mobilization (see Oliver & Marwell, 2001), is facilitated by the quantified, aggregated and visible feedback on group formation embedded in social media platforms. As a result, movement formation dynamics increasingly follow the logic of virality typical of datafied environments, characterized by sudden and unexpected bursts of activity. Acceleration, however, does not necessarily imply a net positive effect on movements organizing, also considering that the same affordances are available to a movement's opponents (Sullivan, 2014). The same mechanism applies to demobilization, which is likewise hastened (see Kwok & Chan, 2022 in this issue). The result might be an acceleration of the 'cycles of contention' (Tarrow, 1998) and the rapid alternation of visible but volatile bursts in contentious activity. The Occupy protest wave is a case in point. Unexpectedly erupting in New York in September 2011, it quickly spread to the rest of the US and in dozens other countries worldwide, to lose suddenly momentum under the pressure of repression and strategic *impasse*. Whereas Occupy Wall Street has been described more as a 'moment' than a 'movement' (Calhoun, 2013), the meta-hashtag #Occupy[+something] has been periodically revived in the following years, contributing to the visibility of local protests worldwide (Beraldo, 2022a).

## Socio-technical political opportunities

Over time, social movement scholarship has paid a great deal of attention to the broader context within which movements emerge. Popular notions include 'political opportunities', or the features of the political-institutional milieu that enable or prevent mobilization (McAdam et al., 1996), as well as 'discursive opportunities', which point to the cultural elements that might facilitate or constrain the attempts to frame an issue for collective action (Koopmans & Statham, 1999). The notion of 'mediation opportunity structure' stresses instead the role of media and communication as instrumental to realizing a movement's goals (Cammaerts, 2012). We note two main developments.

### Data and data infrastructure as socio-technical opportunities

A novel class of political opportunities has emerged, consisting of technology-related disruptions such as data leaks (e.g., Cambridge Analytica), surveillance scandals (e.g., Snowden revelations), or technological advancement that results in the implementation of controversial digital solutions in society (e.g., facial recognition cameras in Hong Kong). These opportunities *change the focus away from the state* and into the corporate sector and the novel state-industry alliances that characterize surveillance capitalism (Zuboff, 2015). Furthermore, empirical evidence exposes how in the digital environment of data 'frames of opportunity appear with far less frequency than threat' (see Wright, 2022 in this issue). In other words, the datafied society in its corporate and state ramifications presents movements with a novel set of challenges that directly impinge on citizen agency and human rights – which may prompt mobilization. We can interpret this evolution as expanding the remit of political opportunities as they have been traditionally defined.

### Novel transnational opportunities

Although sensitive to the local context, datafication creates a matrix for opportunity structures that impact social movements at the global level. This may result in novel opportunity structures that are transnational in nature. Like never before, individuals and communities around the world are united by their use of a set of corporate solutions including devices (e.g., iPhone) and platforms (e.g., Instagram, TikTok) and the 'culture of connectivity' (van Dijck, 2013) they promote, which create a sense of commonality across latitudes, languages, geographies, and cultures (think of how the Arab Spring mobilization have hastily been named 'Twitter revolutions' (El-Nawawy & Khamis, 2012)). This commonality may propel a novel trans-nationalization of issue areas related to the data and technology realm and beyond. The #DeleteFacebook campaign, that swept across the internet in 2018 prompted by a criticism of the company's data manipulation practices, is a case in point (Mills, 2021), as is the 'against Googlization' mobilization in Berlin and Toronto (Charitsis & Laamanen, 2022, in this issue).

### *Data-enabled action repertoires*

Data have entered the movement's action repertoire, adding to the tools available to activists to advance their goals (Beraldo & Milan, 2019). Movements mobilizing on issues like climate change or racism, among others, increasingly embrace data-enabled practices: they may 'start counting' (Currie et al., 2016) to gather independent evidence on social problems for advocacy purposes, as showcased by the examples introducing the article. Novel algorithmic-oriented tactics, such as algorithmic resistance by means of trending topic hijacking (i.e., gaming the rules of visibility of social media platforms) (see Treré & Bonini, 2022, in this issue), doxing (i.e., the online spread of private information about individuals or companies) (see Li & Whitworth, 2023, in this issue), distributed-denial-of-service (DDoS) attacks (i.e., the disruption of a website by overwhelming the service with abnormal traffic) and other 'hacktivist' practices (Jordan, 2015) are increasingly popular among a wide range of movements. Activists might also leverage the technological affordances, such as real-life voting, embedded in technology, such as Telegram (see Poon, 2024, in this issue). We observe three trends.

#### *Novel skill-based partnerships*

Unusual skill-based partnership emerge as claim-making and direct action evolve to exploit the possibilities offered by data and related infrastructure. Innovative coalitions form around newly needed data skills and tactics when these are not readily available to a given movement, with, for example, environmental activists partnering up with data experts and programmers to advance their goals (Milan & Gutiérrez, 2017) or hacker collectives joining in to support of street protesters in disparate contexts (Beraldo, 2022a). This may have the effect of introducing novel divides along the dimension of technological skills.

#### *Loss of control (and increased risk of profiling) due to proprietary data infrastructure*

Data-enabled action repertoires typically rely on corporate platforms whose 'inner' mechanisms (i.e., proprietary algorithms whose workings are not apparent to platform users) are only superficially intelligible to activists, affording scarce control over means and ends of online communication and protest tactics. Movements organizing through Facebook groups, for instance, might suddenly see their structure dissolved due to the platform redefining the boundaries of admissible content or to context-insensitive automated moderation algorithms. And the dependence on proprietary infrastructure is linked with increased risk of activist profiling, online surveillance, and movement-counter movement dynamics (Leistert, 2013).

#### *Repertoire adaptation and 'translation'*

Groups relying on pre-digital tactics **sometimes** seek to identify the 'digital equivalent' of their standard practices, translating them to fit the challenges of the datafied society. The environmental organization Extinction Rebellion offers its activists insights about how to exploit datafied platforms to multiply the visibility of their civil disobedience actions (Extinction Rebellion Present: How To Start A Revolution + Q&A, 2019). The interruption of web-based services of private companies via DDoS attacks, for example, has been compared by activists to acts of expression of free speech such as sit-ins (Karanasiou, 2014). In sum, not only does datafication introduce novel tactics and opportunities, but it may encourage movements to rethink the implications of existing practices considering the datafied environment in which they are deployed.

### *Datafied identity building*

Promoting the reconfiguration of contentious agendas and group formation dynamics, datafication alters the context in which movements enact their subjectivity and negotiate their collective identity. Collective identity denotes the 'common cognitive frameworks' supporting collective action; 'in part the fruit of emotional recognition' (Melucci, 1989, p. 35), it is 'interactive and shared' and results from a 'network of active relationships between actors' (1996, pp. 70–71; c.f. Flesher Fominaya, 2010). Several scholars have sought to re-interpret the concept of collective identity in light of the changing media environment (e.g., Ackland & O'Neil, 2011; Buyukozturk et al., 2018; Kavada, 2015). McDonald noted how collective identity in the digital age approximates 'the public experience of the self', which is inherently fluid, at the cost of group solidarity (2002, p. 111). But how is this process affected by datafication? We identify two trends.

#### *Volatile identities based on platform metrics*

The emergence of contentious subjects in the datafied society is anchored to the logics of the attention economy (Tufekci, 2013) and the grammar of digital culture described above—e.g., the algorithmic configuration of issue publics. Specifically, the 'politics of visibility' (Milan, 2015b) prompted by the personal engagement with content production and platform metrics, centers the individual's subjective and private experience, partly replacing the formation of a collective identity resilient to time. It makes it susceptible to continuous renegotiation, amplifying its 'interactive and shared' character, to revisit Melucci (Milan, 2015b) – however with the intervention of algorithms outside movement control. Collective identity then tends to be volatile and short-lived, but it is also flexible to accommodate multiple affiliations. Hashtag activism is a prime example of this phenomenon: the #MeToo marker connected personal experiences of sexual harassment allowed for 'network acknowledgement' (Suk et al., 2021), exposing how magnitude is a mechanism for validation of experiences and identities (Gerbaudo, 2022). This validation is however only possible based on minimum-denominator descriptors (Milan, 2015a).

#### *Incoherence and exacerbation of a movement's internal diversity*

Contentious instances, as seen in relation to group formation, are reassembled by algorithmic processes into constellations that are contingent, temporary, and (potentially) incoherent. Their fluid and transient nature can be masked by the bold, stable appearance of their semiotic interface, as social media platforms encourage the adoption of a standardized symbolism to partake in their 'visibility game'. Recognizable 'brands' propelled by datafied measures of popularity (e.g., Anonymous, #Occupy, Black Lives Matters) represent the same, uniform face of diverse, and often incoherent assemblages (Beraldo, 2020, 2022a). If contradiction in movement formations is nothing new, as movement identities are naturally open for negotiation and internal conflict (Melucci, 1996), this tendency has been exacerbated by the personalization of contentious politics favored by digital communication networks (Bennett & Segerberg, 2012) and, moreover, by the platform dynamics implicit in datafied identity formation.

### *Machine-mediated meaning work*

Meaning-making, or the activists' 'interactive process of constructing meaning' for collective action (Gamson, 1992, p. xii), is at the core of movements' activity (see also, inter alia, Melucci, 1996; Mische, 2003; Polletta, 2006). Because 'issues do not have an independent life outside of people's efforts to characterize them as such' (della Porta & Diani, 2006, p. 65), it matters how movement actors assign meaning to their experience and whom they interact with to elaborate their definitions of issues (e.g., air pollution) to link them to a broader problem (e.g., climate change). The changing technological landscape affects the way meaning work unfolds within and around movements, primarily by changing the ways its material basis (i.e., information) is generated, distributed and consumed. Meaning work today is increasingly mediated by algorithms and AI-powered search engines and recommendation systems, including consumer-level Large Language Models (LLM), either as standalone applications or integrated into others. With two main outcomes:

#### *Hybrid and biased info-scape*

Because people's 'information diet' is at the core of public opinion and values formation, the dynamics related to the production, circulation and consumption of information are constitutive of social movements in contemporary societies (see Castells, 2009). With the advent of datafication, these dynamics are increasingly the outcome of hybrid processes, where the human and the machine are intertwined. Algorithms filter, rank and rearrange information for users, allegedly augmenting political polarization, either by trapping users in echo chambers of homogeneous content, or by exposing them to positions outside of their bubble and reinforcing partisanship as the result of group interaction dynamics (Törnberg, 2022). And LLMs intervene not only at the level of curation, but also at the level of generation of information, further hybridizing the contribution of humans and machines in defining today's info-scape. In other words, AI-operated agents become part of the constellation of players that contribute to the 'sustained symbolic and cultural conflict between different actors' (della Porta & Diani, 2006, p. 66): an add-on to the interpretative tools and cognitive apparatus (made of 'cultural and ideational elements') that enable people to make sense of their world (Swidler, 1986, p. 273). Whereas it is still early to appreciate the profound transformations that 'Artificial Intelligence' capable of imitating the autonomy and creativity of human actors will bring about, these technologies stand in continuity with less advanced forms of algorithmic mediation of information in that they are characterized by intrinsic bias (cf., Kordzadeh & Ghasemaghaei, 2022) – a condition that contemporary social movements can no longer ignore and, occasionally, seek to resist (see the case of algorithmic transparency activism).

### *Data as rhetorical tools for framing*

Cognitive praxis, e.g., the relentless production of cultural codes (Melucci, 1989), is a constitutive element of organized collective action. Today data are mobilized as part of the narrative strategy of certain social movements and, more broadly, as resources for their framing activity (see Benford & Snow, 2000). The climate justice movement is a case in point, as it oftentimes relies on statistics and data visualizations as tools to construct its diagnostic framing (cf., Vera et al., 2019). Data-solutionist movements such as 'data science for the social good' initiatives (e.g., Kinsella, 2021) incorporate data as part of their prognostic framing, while the emphasis on social media engagement metrics as a vehicle for recruitment and to demonstrate societal relevance testifies of how datafication can impact motivational framing.

## Conclusions

This article asked how the 'movement society' (Meyer & Tarrow, 1998) is responding to the advent and challenges of datafication. We explored the tensions that big data (and, by extension, up-and-coming Artificial Intelligence applications) expose social movement research to. We started charting the building block of an emerging theory of datafied movements that brings the discipline in conversation with critical data studies. We argued that to understand social movement dynamics in the age of datafication we cannot look at data and data infrastructure in a merely instrumental way, e.g., as a methodological toolkit ('big data research'), nor can we consider data solely in its guise of 'information'. This is because datafication represents a fundamental paradigm shift with consequences on the way we exercise and understand citizenship, political participation, and collective action. It is also a totalizing experience that can hardly be avoided, as it deeply affects the context in which movements emerge and operate. Its consequences for social movements go well beyond what is captured by notions such as digital or data activism. To meet these challenges, we advocated for attending to the infrastructural and material dimension of the relation between movements and data, embracing a socio-technical perspective in studying datafied movements. We advanced the notion of 'datafied movements' precisely to capture these developments. Rather than indicating a novel class of activism, datafied movements stand in for the structural condition of contentious politics in the age of datafication: in other words, we are looking at a novel socio-historical configuration of the social movement society. Hence, we revisited five key elements of the conceptual toolbox of the discipline, namely group formation, opportunity structures, action repertoires, meaning work, and collective identity, in the assumption that social movement scholarship needs a makeover to fully understand the tensions within datafied movements. Algorithmic group making, socio-technical political opportunities, data-enabled action repertoires, datafied identity building, and machine-mediated meaning work meet the demand for citizen participation today, presenting social movement action with novel challenges. More work remains to be done to future-proof social movement studies for the complex technological developments we live through. We thus encourage colleagues to pick up the challenge of better understanding the role of data and data infrastructure in re-mediating social movements dynamics amidst rapid technological innovation.

## Note

1. Furthermore, movements might actively avoid or seek to minimize the effects of datafication on their mobilization strategies, withdrawing from the 'matrix' of datafication. But because datafication cannot be ignored but resisted, this case, too, exposes the grasp of datafication on social actors. Movements, then, will adapt their practices to minimize its reach, continuously re-negotiating their existence in an increasingly datafied society.

## Disclosure statement

No potential conflict of interest was reported by the author(s).

## Funding

This research was made possible by funding from the European Research Council (ERC) under the European Union's Horizon 2020 research and innovation program [grant agreement No. 639379-DATACTIVE], awarded to Stefania Milan as Principal Investigator.

## ORCID

Stefania Milan http://orcid.org/0000-0002-9314-2889
Davide Beraldo http://orcid.org/0000-0002-8501-596X

## References

Ackland, R., & O'Neil, M. (2011). Online collective identity: The case of the environmental movement. *Social Networks*, *33*(3), 177–190. https://doi.org/10.1016/j.socnet.2011.03.001

Atton, C. (2004). *An alternative internet: Radical media, politics and creativity*. Edinburgh University Press.

Barassi, V., & Zamponi, L. (2020). Social media time, identity narratives and the construction of political biographies. *Social Movement Studies*, *19*(5–6), 592–608. https://doi.org/10.1080/14742837.2020.1718489

Barreneche, C., & Lombana-Bermudez, A. (2023). Another infrastructure is possible: Grassroots citizen sensing and environmental data justice in Colombia. *International Journal of Communication*, *17*, 3638–3659. https://ijoc.org/index.php/ijoc/article/view/18821/4193

Beer, D. (2016). *Metric power*. Palgrave MacMillan.

Benford, R. D., & Snow, D. A. (2000). Framing processes and social movements: An overview and assessment. *Annual Review of Sociology*, *26*(1), 611–639. https://doi.org/10.1146/annurev.soc.26.1.611

Bennett, L. W., & Segerberg, A. (2012). The logic of connective action: Digital media and the personalization of contentious politics. *Information, Communication & Society*, *15*(5), 739–768. https://doi.org/10.1080/1369118X.2012.670661

Beraldo, D. (2020). Dividing by zero. Tautology and paradox in the self-descriptions of anonymous. *Journal of Sociocybernetics*, *17*(1). https://doi.org/10.26754/ojs_jos/jos.202014246

Beraldo, D. (2022a). Movements as multiplicities and contentious branding: Lessons from the digital exploration of #Occupy and #Anonymous. *Information, Communication & Society*, *25*(8), 1098–1114. https://doi.org/10.1080/1369118X.2020.1847164

Beraldo, D. (2022b). Unfolding #Anonymous on twitter. The networks behind the mask. *First Monday*, *27*(1–3). https://doi.org/10.5210/fm.v27i1.11723

Beraldo, D., & Milan, S. (2019). From data politics to the contentious politics of data. *Big Data & Society*, *6*(2), 1–11. https://doi.org/10.1177/2053951719885967

Bimber, B., Flanagin, A., & Stohl, C. (2012). *Collective action in organizations: Interaction and engagement in an era of technological change*. Cambridge University Press.

Buyukozturk, B., Gaulden, S., & Dowd-Arrow, B. (2018). Contestation on reddit, gamergate, and movement barriers. *Social Movement Studies*, *17*(5), 592–609. https://doi.org/10.1080/14742837.2018.1483227

Calhoun, C. (2013). Occupy wall street in perspective. *The British Journal of Sociology*, *64*(1), 26–38. https://doi.org/10.1111/1468-4446.12002

Cammaerts, B. (2012). Protest logics and the mediation opportunity structure. *European Journal of Communication*, *27*(2), 117–134. https://doi.org/10.1177/0267323112441007

Cammaerts, B., Mattoni, A., & McCurdy, P. (Eds.). (2013). *Mediation and protest movements*. Intellect.

Carroll, W. K., & Hackett, R. A. (2006). Democratic media activism through the lens of social movement theory. *Media, Culture & Society*, *28*(1), 83–104. https://doi.org/10.1177/0163443706059289

Castells, M. (2009). *Communication power*. Oxford University Press.

Charitsis, V., & Laamanen, M. (2022). When digital capitalism takes (on) the neighbourhood: Data activism meets place-based collective action. *Social Movement Studies*, 1–18. https://doi.org/10.1080/14742837.2022.2123314

Costanza-Chock, S. (2012). Mic check! Media cultures and the occupy movement. *Social Movement Studies: Journal of Social, Cultural and Political Protest*, *11*(3–4), 375–385. https://doi.org/10.1080/14742837.2012.710746

Couldry, N. (2020). Recovering critique in an age of datafication. *New Media & Society*, *22*(7), 1135–1151. https://doi.org/10.1177/1461444820912536

Currie, M., Paris, B., Pasquetto, I., & Pierre, J. (2016). The conundrum of police officer-involved homicides: Counter-data in Los Angeles County. *Big Data & Society*, *3*(2), 205395171666356. https://doi.org/10.1177/2053951716663566

Data for Black Lives. (n.d.). *About data for black lives*. http://d4bl.org/about.html

della Porta, D., & Diani, M. (2006). *Social movements. An introduction* (2nd ed.). Blackwell.

Della Porta, D., & Pavan, E. (2017). Repertoires of knowledge practices: Social movements in times of crisis. *Qualitative Research in Organizations and Management*, *12*(4), 297–314. https://doi.org/10.1108/QROM-01-2017-1483

Diani, M., & Mische, A. (2015). Network approaches and social movements. In D. D. Porta & M. Diani (Eds.), *The oxford handbook of social movements* (pp. 306–325). Oxford University Press.

Dolata, U., & Schrape, J.-F. (2016). Masses, crowds, communities, movements: Collective action in the Internet age. *Social Movement Studies*, *15*(1), 1–18. https://doi.org/10.1080/14742837.2015.1055722

Dunbar-Hester, C. (2014). *Low power to the people! Pirates, protest, and politics in FM radio activism*. MIT Press.

El-Nawawy, M., & Khamis, S. (2012). Political activism 2.0: Comparing the role of social media in Egypt's "facebook revolution" and Iran's "twitter uprising". *CyberOrient*, *6*(1), 8–33. https://doi.org/10.1002/j.cyo2.20120601.0002

Eubanks, V. (2018). *Automating inequality: How high-tech tools profile, police, and punish the poor*. St. Martin's Press.

*Extinction Rebellion Present: How To Start A Revolution + Q&A*. (2019, August 27). https://www.youtube.com/watch?v=ngBMg-jKaG4

Ferguson, A. G. (2017). *The rise of big data policing: Surveillance, race, and the future of law enforcement*. New York University Press.

Flesher Fominaya, C. (2010). Collective identity in social movements: Central concepts and debates. *Sociology Compass*, *4*(6), 393–404. https://doi.org/10.1111/j.1751-9020.2010.00287.x

Flesher Fominaya, C., & Gillan, K. (2017). Navigating the technology-media-movements complex. *Social Movement Studies*, *16*(4), 383–402. https://doi.org/10.1080/14742837.2017.1338943

Gamson, W. A. (1992). *Talking politics*. Cambridge University Press.

Garrett, K. (2006). Protest in an information society: A review of literature on social movements and new ICTs. *Information, Communication & Society*, *9*(2), 202–224. https://doi.org/10.1080/13691180600630773

Gerbaudo, P. (2012). *Tweets and the streets. Social media and contemporary activism*. Pluto Press.

Gerbaudo, P. (2022). From individual affectedness to collective identity: Personal testimony campaigns on social media and the logic of collection. *New Media & Society*, 14614448221128523. https://doi.org/10.1177/14614448221128523

Gillespie, T. (2014). The relevance of algorithms. In T. Gillespie, P. Boczkowski, & K. Foot (Eds.), *Media technologies: Essays on communication, materiality, and society* (pp. 167–194). MIT Press.

González-Bailón, S., Borge-Holthoefer, J., Rivero, A., & Moreno, Y. (2011). The dynamics of protest recruitment through an online network. *Scientific Reports*, *1*(1), 197. https://doi.org/10.1038/srep00197

Gutiérrez, M., & Milan, S. (2019). Playing with data and its consequences. *First Monday*, *24*(1). https://doi.org/10.5210/fm.v24i1.9554

Hardt, M. (2017). Multiple temporalities of the movements. *tripleC: Communication, capitalism & Critique. tripleC: Communication, capitalism & critique. Open Access Journal for a Global Sustainable Information Society*, *15*(2), 390–392. https://doi.org/10.31269/triplec.v15i2.888

Heawood, J. (2018). Pseudo-public political speech: Democratic implications of the Cambridgeanalytica scandal. *Information Polity*, *23*(4), 429–434. https://doi.org/10.3233/IP-180009

Heise, U. K. (2002). Unnatural ecologies: The metaphor of the environment in media theory. *Configurations*, *10*(1), 149–168. https://doi.org/10.1353/con.2003.0006

Hess, D. J. (2016). *Undone science. Social movements, mobilized publics, and industrial transitions*. MIT Press.

Hintz, A., & Brown, I. (2017). Enabling digital citizenship? The reshaping of surveillance policy after Snowden. *International Journal of Communication*, *11*, 782–801. https://ijoc.org/index.php/ijoc/article/view/5522/1931

Hintz, A., Dencik, L., & Wahl-Jorgensen, K. (2018). *Digital citizenship in a datafied society*. Polity.

Iliadis, A., & Russo, F. (2016). Critical data studies: An introduction. *Big Data & Society*, *3*(2), 2053951716674238. https://doi.org/10.1177/2053951716674238

Jasanoff, S. (2017). Virtual, visible, and actionable: Data assemblages and the sightlines of justice. *Big Data & Society*, *4*(2), 205395171772447. https://doi.org/10.1177/2053951717724477

Jordan, T. (2015). *Information politics. Liberation and exploitation in the digital society*. Pluto Press.

Karanasiou, A. P. (2014). The changing face of protests in the digital age: On occupying cyberspace and distributed-denial-of-services (DDoS) attacks. *International Review of Law, Computers and Technology*, *28*(1), 98–113. https://doi.org/10.1080/13600869.2014.870638

Kaun, A., & Uldam, J. (2018). Digital activism: After the hype. *New Media & Society*, *20*(6), 2099–2106. https://doi.org/10.1177/1461444817731924

Kavada, A. (2015). Creating the collective: Social media, the occupy movement and its constitution as a collective actor. *Information, Communication & Society*, *18*(8), 872–886. https://doi.org/10.1080/1369118X.2015.1043318

Kavada, A. (2016). Social movements and political agency in the digital age: A communication approach. *Media and Communication*, *4*(4), 8–12. https://doi.org/10.17645/mac.v4i4.691

Kazansky, B. (2021). 'It depends on your threat model': The anticipatory dimensions of resistance to data-driven surveillance. *Big Data & Society*, *8*(1), 2053951720985557. https://doi.org/10.1177/2053951720985557

Keck, M., & Sikkink, K. (1998). *Activists beyond borders. Advocacy networks in international policy*. Cornell University Press.

Kinsella, B. (2021). Data science for social good volunteer motivations and limitations: An exploratory survey. *Chance*, *34*(3), W86–W95. https://doi.org/10.1080/09332480.2021.1981055

Kitchin, R. (2014). Big Data, new epistemologies and paradigm shifts. *Big Data & Society*, *1*(1), 205395171452848. https://doi.org/10.1177/2053951714528481

Koopmans, R., & Statham, P. (1999). Political claims analysis: Integrating protest event and political discourse approaches. *Mobilization: An International Quarterly*, *4*(2), 203–221. https://doi.org/10.17813/maiq.4.2.d7593370607l6756

Kordzadeh, N., & Ghasemaghaei, M. (2022). Algorithmic bias: Review, synthesis, and future research directions. *European Journal of Information Systems*, *31*(3), 388–409. https://doi.org/10.1080/0960085X.2021.1927212

Kwok, C., & Chan, N. K. (2022). Datafication and implicated networks of demobilization: Social movement demobilization in datafied societies. *Social Movement Studies*, 1–18. https://doi.org/10.1080/14742837.2022.2128326

Latour, B. (2005). *Reassembling the social: An introduction to actor-network-theory*. Oxford University Press. http://dss-edit.com/plu/Latour_Reassembling.pdf

Leistert, O. (2013). *From protest to surveillance—the political rationality of mobile media. Modalities of neoliberalism*. Peter Lang.

Li, Y.-T., & Whitworth, K. (2023). Coordinating and doxing data: Hong Kong protesters' and government supporters' data strategies in the age of datafication. *Social Movement Studies*, 1–18. https://doi.org/10.1080/14742837.2023.2178404

López, J. (2020). *Experimentando con la pobreza: El Sisbén y los proyectos de analítica de datos*. Fundación Karisma. https://web.karisma.org.co/wp-content/uploads/download-manager-files/Experimentando%20con%20la%20pobreza.pdf

Madison, N., & Klang, M. (2019). Recognizing Everyday Activism: Understanding Resistance to Facial Recognition. *Journal of Resistance Studies*, *2*(5), 97–113.

Mattoni, A., & Pavan, E. (2018). Politics, participation and big data. Introductory reflections on the ontological, epistemological, and methodological aspects of a complex relationship. *Participazione & Conflitto*, *11*(2), 313–331. https://doi.org/10.1285/i20356609v11i2p313

Mattoni, A., & Treré, E. (2014). Media practices, mediation processes, and mediatization in the study of social movements. *Communication Theory*, *24*(3), 252–271. https://doi.org/10.1111/comt.12038

McAdam, D., McCarthy, J. D., & Zald, M. N. (Eds.). (1996). *Comparative perspectives on social movements: Political opportunities, mobilizing structures, and cultural framings*. Cambridge University Press.

Meikle, G. (2002). *Future active: Media activism and the internet*. Routledge.

Mejias, U. A., & Couldry, N. (2019). Datafication. *Internet Policy Review*, *8*(4). https://doi.org/10.14763/2019.4.1428

Melucci, A. (1989). *Nomads of the present. Social movements and individual needs in contemporary society*. Hutchinson Radius.

Melucci, A. (1996). *Challenging codes. Collective action in the information age*. Cambridge University Press.

Meng, A., & DiSalvo, C. (2018). Grassroots resource mobilization through counter-data action. *Big Data & Society*, *5*(2), 2053951718796862. https://doi.org/10.1177/2053951718796862

Meyer, D. S., & Tarrow, S. (Eds.). (1998). *The social movement society. Contentious politics for a new century*. Rowman & Littlefield.

Milan, S. (2013). *Social movements and their technologies: Wiring social change*. Palgrave Macmillan.

Milan, S. (2015a). From social movements to cloud protesting: The evolution of collective identity. *Information, Communication & Society*, *18*(8), 887–900. https://doi.org/10.1080/1369118X.2015.1043135

Milan, S. (2015b). When algorithms shape collective action: Social media and the dynamics of cloud protesting. *Social Media + Society*, *1*(2), 205630511562248. https://doi.org/10.1177/2056305115622481

Milan, S., & Barbosa, S. (2020). Enter the WhatsApper: Reinventing digital activism at the time of chat apps. *First Monday*, *25*(1). https://doi.org/10.5210/fm.v25i12.10414

Milan, S., & Gutiérrez, M. (2017). Technopolitics in the Age of Big Data. In F. S. Caballero & T. Gravante (Eds.), *Networks, Movements & Technopolitics in Latin America: Critical Analysis and Current Challenges* (pp. 95–109). Palgrave MacMillan.

Milan, S., Treré, E., & Masiero, S. (2021). *COVID-19 from the margins: Pandemic invisibilities, policies and resistance in the datafied society*. Institute of Network Cultures. https://networkcultures.org/wp-content/uploads/2021/02/Covid19FromTheMargins-1.pdf

Milan, S., & van der Velden, L. (2016). The alternative epistemologies of data activism. *Digital Culture & Society*, *2*(2), 57–74. https://doi.org/10.14361/dcs-2016-0205

Mills, S. (2021). #deletefacebook: From popular protest to a new model of platform capitalism? *New Political Economy*, *26*(5), 851–868. https://doi.org/10.1080/13563467.2020.1858777

Mische, A. (2003). Cross-talk in movements: Reconceiving the culture-network link. In M. Diani & D. McAdam (Eds.), *Social movements and networks. Relational approaches to collective action* (pp. 258–280). Oxford University Press.

Mortesen, M., Neumayer, C., & Poell, T. (Eds.). (2019). *Social media materialities and protest: Critical reflections*. Routledge.

Muravyov, D. (2022). Doubt to be certain: Epistemological ambiguity of data in the case of grassroots mapping of traffic accidents in Russia. *Social Movement Studies*, 1–17. https://doi.org/10.1080/14742837.2022.2128327

Neumayer, C., & Rossi, L. (2016). 15 years of protest and media technologies scholarship: A sociotechnical timeline. *Social Media + Society*, *2*(3), 2056305116662180. https://doi.org/10.1177/2056305116662180

Noakes, J. A., & Johnston, H. (2005). Frames of protest: A roadmap to a perspective. In H. Johnston & J. A. Noakes (Eds.), *Frames of protest: Social movements and the framing perspective* (pp. 1–30). Rowman & Littlefield.

Oliver, P. E., & Marwell, G. (2001). Whatever happened to critical mass theory? A retrospective and assessment. *Sociological Theory*, *19*(3), 292–311. https://doi.org/10.1111/0735-2751.00142

Papacharissi, Z. (2010). *A private self: Democracy in a digital age*. Polity Press.

Pickard, V. (2006). United yet autonomous: Indymedia and the struggle to sustain a radical democratic network. *Media, Culture & Society*, *28*(3), 315–336. https://doi.org/10.1177/0163443706061685

Pickard, V., & Yang, G. (Eds.). (2017). *Media activism in the digital age*. Routledge.

Poell, T. (2020). Social media, temporality, and the legitimacy of protest. *Social Movement Studies*, *19*(5–6), 609–624. https://doi.org/10.1080/14742837.2019.1605287

Poell, T., Abdulla, R., Rieder, B., Woltering, R., & Zack, L. (2016). Protest leadership in the age of social media. *Information, Communication & Society*, *19*(7), 994–1014. https://doi.org/10.1080/1369118X.2015.1088049

Polletta, F. (2006). *It was like a fever: Storytelling in protest and politics*. University of Chicago Press.

Poon, M. (2024). Achieving organizationality through authorship affordances — a communicative episode of telegram polling from 2019 Hong Kong. *Social Movement Studies*, 1–22. https://doi.org/10.1080/14742837.2023.2296868

Rajão, R., & Jarke, J. (2018). The materiality of data transparency and the (re)configuration of environmental activism in the Brazilian Amazon. *Social Movement Studies*, *17*(3), 318–332. https://doi.org/10.1080/14742837.2018.1459297

Ruppert, E., Isin, E., & Bigo, D. (2017). Data Politics. *Big Data & Society*, *4*(2), 1–7. https://doi.org/10.1177/2053951717717749

Ryan, C., & Jeffreys, K. (2019). *Beyond prime time activism. Communication activism and social change.* Routledge.

Snow, D. A. (2004). Framing process, ideology, and discursive fields. In D. A. Snow, S. A. Soule, & H. Kriesi (Eds.), *The blackwell companion to social movements* (pp. 380–412). Blackwell.

Srnicek, N. (2016). *Platform capitalism.* Polity Press.

Suk, J., Abhishek, A., Zhang, Y., Ahn, S. Y., Correa, T., Garlough, C., & Shah, D. V. (2021). #metoo, networked acknowledgment, and connective action: How "empowerment through empathy" launched a social movement. *Social Science Computer Review*, *39*(2), 276–294. https://doi.org/10.1177/0894439319864882

Sullivan, J. (2014). China's Weibo: Is faster different? *New Media & Society*, *16*(1), 24–37. https://doi.org/10.1177/1461444812472966

Sun, Y., & Yin, S. (2022). Opening up mediation opportunities by engaging grassroots data: Adaptive and resilient feminist data activism in China. *New Media & Society*, 146144482210968. https://doi.org/10.1177/14614448221096806

Swidler, A. (1986). Culture in action: Symbols and Strategies. *American Sociological Review*, *51*(2), 273–286. https://doi.org/10.2307/2095521

Tarrow, S. (1998). *Power in movement. Social movements and contentious politics.* Cambridge University.

Törnberg, P. (2022). How digital media drive affective polarization through partisan sorting. *Proceedings of the National Academy of Sciences*, *119*(42), e2207159119. https://doi.org/10.1073/pnas.2207159119

Torres, G. (2019). Institutional resistance to transparency: The quest for public sector information in Mexico. *Journal of Resistance Studies*, *5*(2), 68–96.

Treré, E. (2016). The dark side of digital politics: Understanding the algorithmic manufacturing of consent and the hindering of online dissidence. *IDS Bulletin*, *41*(1), 127–138. https://doi.org/10.19088/1968-2016.111

Treré, E. (2019). *Hybrid media activism.* Routledge.

Treré, E., & Bonini, T. (2022). Amplification, evasion, hijacking: Algorithms as repertoire for social movements and the struggle for visibility. *Social Movement Studies*, 1–17. https://doi.org/10.1080/14742837.2022.2143345

Trottier, D., & Fuchs, C. (Eds.). (2014). *Social media, politics and the state: Protests, revolutions, riots, crime and policing in the age of Facebook, Twitter and YouTube.* Routledge.

Tufekci, Z. (2011). *Faster is different.* Technosociology. Our Tools, Ourselves. http://technosociology.org/?p=424

Tufekci, Z. (2013). "Not this one": Social movements, the attention economy, and microcelebrity networked activism. *American Behavioral Scientist*, *57*(7), 848–870. https://doi.org/10.1177/0002764213479369

van Dijck, J. (2013). *The culture of connectivity: A critical history of social media.* Oxford University Press.

van Dijck, J. (2014). Datafication, dataism and dataveillance: Big data between scientific paradigm and ideology. *Surveillance and Society*, *12*(2), 197–208. https://doi.org/10.24908/ss.v12i2.4776

van Dijck, J., & Poell, T. (2013). Understanding social media logic. *Media and Communication*, *1*(1), 2–14. https://doi.org/10.17645/mac.v1i1.70

van Es, K., & Schäfer, M. T. (Eds.). (2017). Introduction. New brave world. In *The datafied society. Studying culture through data* (pp. 13–22). Amsterdam University Press.

Vera, L. A., Dillon, L., Wylie, S., Ohayon, J. L., Lemelin, A., Brown, P., Sellers, C., Walker, D., & the Environmental Data and Governance Initiative. (2018). Data resistance: A social movement organizational autoethnography of the environmental data and governance initiative. *Mobilization: An International Quarterly*, *23*(4), 511–529.

Vera, L. A., Walker, D., Murphy, M., Mansfield, B., Siad, L. M., & Ogden, J. (2019). When data justice and environmental justice meet: Formulating a response to extractive logic through environmental data justice. *Information, Communication & Society*, *22*(7), 1012–1028. https://doi.org/10.1080/1369118X.2019.1596293

Whittaker, J., Looney, S., Reed, A., & Votta, F. (2021). Recommender systems and the amplification of extremist content. *Internet Policy Review*, *10*(2). https://doi.org/10.14763/2021.2.1565

Wissenbach, K. R. (2020). Accounting for power in transnational civic tech activism: A communication-based analytical framework for media practice. *International Communication Gazette*, *82*(6), 545–563. https://doi.org/10.1177/1748048519832779

Wong, J. (2020). Protests decentralised: How technology enabled civil disobedience by Hong Kong anti-extradition bill protesters. *8th Asian Privacy Scholars Network Conference*. https://doi.org/10.31228/osf.io/efvwn

Wright, J. M. (2022). "The future of the internet hangs in the balance": The perception and framing of political opportunity and threat in the contentious politics of data. *Social Movement Studies*, 1–18. https://doi.org/10.1080/14742837.2022.2123313

Zalnieriute, M. (2021). Protests and public space surveillance: From metadata tracking to facial recognition technologies. *Thematic Report to the 50th Session of the UN Human Rights Council*. https://doi.org/10.2139/ssrn.3882317

Zuboff, S. (2015). Big other: Surveillance capitalism and the prospects of an information civilization. *Journal of Information Technology*, *30*(1), 75–89. https://doi.org/10.1057/jit.2015.5

# "The future of the internet hangs in the balance": the perception and framing of political opportunity and threat in the contentious politics of data

Jared M. Wright

**ABSTRACT**

Opportunities are often conceptualized as 'openings' in a social movement's external environment which can reduce the cost of collective action, while threat is a force which increases the cost of inaction. These concepts were originally formulated to describe the political contexts of traditional offline movements, therefore how are they perceived and framed by activists in the digital environment of data? This study utilizes qualitative thematic coding to examine the collective action frames in four years of archival texts from 2009 to 2012 of two highly adept data activist mobilizations: The Digital Rights movement and the Anonymous hacktivist collective. Analyses show that frames of opportunity appear with far less frequency than threat. For both cases, volatile opportunity frames regarding their online actions are very rare, especially in the Anonymous texts. It concludes by suggesting that highly leveraging the affordances of digital technology and data may lead to political opportunities losing some of their perceived salience as a mobilizing force, while cultural factors around data activism may simultaneously result in an increased perception of threats. This study shows the importance of understanding how data mediate contemporary collective action and calls for further development and synthesis of the structural and cultural aspects of both opportunity and threat in social movement theory.

## Introduction

Research has pointed to the emergence of a 'social movement society' (Meyer & Tarrow, 1998; Park & Einwohner, 2019; Soule & Earl, 2005) in which collective action and protest have become increasingly accepted forms of political participation in democratic societies, particularly since the 1960s. In this way, social movements have come to constitute a crucial part of the public sphere (Habermas, 1989). Likewise, scholars across multiple disciplines have explored how the Internet has helped form new 'transnational' (Guidry et al., 2000) or 'networked' (Castells, 2015; Shirky, 2011; Tufekci, 2017) public spheres by providing digital spaces where people can discuss, debate, and collectively organize. In doing so, the Internet has helped to further propel the movement society by enabling larger numbers of even more diverse claims across the political spectrum (Earl & Kimport, 2009).

Yet, the Internet is also a highly contested space. Leaders in the technology industry have optimistically described it as, 'the largest experiment involving anarchy in history . . . the world's largest ungoverned space' (Schmidt & Cohen, 2014, p. 3). However, states continually seek ways to surveil, censor, and control the digital flow of information (Greenwald, 2014; Hu, 2017; Wills, 2017). Correspondingly, information has become the new capital (Johnson Andrews, 2019), and corporations increasingly work to privatize and monetize information through datafication under the new economic logic of 'surveillance capitalism' (Van Dijck, 2014; Zuboff, 2019). These corporate entities in turn fuse with the modern security state to form new techno-legal regimes in which secretive arrangements of technology and data sharing enable unparalleled panoptic surveillance and information control (Wills, 2017; Zuboff, 2019).

In response, people across the world have collectively organized around technological data practices and communities to protest this imposition of control (Beraldo & Milan, 2019; Dencik et al., 2016; Milan & van der Velden, 2016). While much research has focused on how a wide variety of contemporary mobilizations leverage digital affordances (Bennett & Segerberg, 2013; Earl & Kimport, 2011; Gerbaudo, 2012; Milan, 2015a, 2015b; Schradie, 2019; Treré, 2019; Tufekci, 2017), relatively few have examined the movements which critically interrogate the technologies themselves and resist how they are being used against individuals by techno-legal regimes of power. Beraldo and Milan (2019) call this the *contentious politics of data*, or 'the multiplicity of bottom-up, transformative initiatives interfering with and/or hijacking dominant, top-down processes of datafication, by contesting existing power relations and narratives and/or by re-appropriating data practices and infrastructure for purposes distinct from the intended' (p. 2). Data fundamentally mediate such activism, also known as *data activism*, either as the 'stakes' of action or the 'repertoire' of tactics (p. 8).

Building upon this analytical framework, this study asks how data might mediate other collective action dynamics; specifically, the perception and framing of political opportunities and threats. Opportunities are typically conceptualized as structural 'openings' in a movement's external environment which can reduce the cost of engaging in collective action (Tarrow, 1998). Originally, this concept was developed to explain the political environment of traditional offline movements (McAdam, 1982). But how might political opportunity structures, and the relatively newer concept of threat, be perceived differently in the environment of digital data? This study answers those questions by analyzing collective action frames in four years (2009–2012) of newsletters and press releases from two cases of contentious data politics fighting for what they see as the 'Freedom of the Internet': The Digital Rights movement (Daskal, 2018; Postigo, 2012) and the Anonymous hacktivist collective (Coleman, 2014).

This study will contribute to social movement scholarship in several ways. First, these cases are compelling because they combine both dimensions of data activism by fighting to subvert datafication and information control (data as stakes) while also utilizing the most innovative cutting-edge data practices and tactics (data as repertoire). Second, examining these activists can add theoretically important insights into changing collective action dynamics resulting from these relatively new, developing technological practices. These activists may even foreshadow changes that will one day affect all social movements. What is currently seen as technological expertise in data activism will likely become far more normative in coming years as datafication becomes more pervasive and the significance of digital space as a site of protest continues to grow. Third, this ongoing contention over issues like encryption and information control is actively shaping the infrastructure of digital space itself, the outcomes of which may ultimately have profound impacts on everyone's online privacy and freedom of expression (DeNardis, 2012; Fish & Follis, 2016; Milan & Ten Oever, 2017). Understanding this phenomenon should therefore be of great interest to all people living in the digital age whose lives are increasingly intertwined with technology and data.

I begin by describing the cases and their theoretical significance, followed by a review of the relevant literature on political opportunity and threat. After explaining the data selection and analysis, and then present my key findings with selected passages from the texts. I conclude with a discussion of how data can mediate the structural and cultural factors that determine how opportunities and threats are perceived and framed in contemporary activism and call for a more unified theoretical framework for these concepts.

## Data activism for internet freedom

There have been numerous campaigns in recent years reflecting an emerging sense of digital values regarding how people use the Internet and control their data. My study design focuses on two such cases, the Digital Rights movement (DR) and the Anonymous hacktivist collective (ANON). Each can be thought of as independent social movements in that they are networks of individuals or groups engaged in sustained episodes of contention based on a shared collective identity (Diani & Bison, 2008). Both are heavily influenced by online 'hacker' culture (Levy, 1984) and arose as direct responses to the techno-legal regimes and institutions which seek to privatize and control digital space and data (Coleman, 2012, 2014, 2017; Daskal, 2018; Postigo, 2012). As such, these activists are best positioned to leverage the key affordances of digital technology and data while simultaneously working to promote what they see as digital rights and freedoms.

These cases stand out because technologies function as more than just instrumental tools. Their design practices are explicitly political; therefore, the technologies are inherently meaningful. As Postigo (2012) explains, 'Their creation and existence can be read as a form of protest (so they are meaningful beyond their function), but they also realize part of the central goal the movement seeks: a culture that is participatory (with the tools to engage in participation)' (p. 8). Furthermore, digital technologies comprise both the obstacles that activists face and the means they use to resist. Lessig (2006) argues that while states traditionally use law as the primary means of social control, with technology they can now control behavior directly in extra-legal ways. This is known as 'technological enforcement,' as is visible in data practices such as copyright controls, online censorship, dataveillance, and algorithmic governance. 'The Digital Rights movement, unlike many other social movements, confronts not only legal regimes, but technological regimes as well, some of which exist outside the reach of traditional institutional mechanisms for social change' (Postigo, 2012, p. 11).

The cases in this study can be seen as counter-hegemonic movements using data to resist these techno-legal regimes. Central to each of these movements are contemporary hacker communities which have been mobilized by external events to become political players, converting their coding tools into what Coleman (2017) calls 'weapons of the geek.'[1] Accounting for both affordances *and* practices (Flesher Fominaya & Gillan, 2017), such activists should be most capable of leveraging the affordances of digital technology and data to their full advantage, which can lower the costs of engaging in collective action (Bennett & Segerberg, 2013; Earl & Kimport, 2011; Highfield, 2016). They also work to create 'counter-imaginaries' through their data practices to subvert dominant cultural views of processes like datafication as beneficial and inevitable (Kazansky & Milan, 2021). In doing so, they are not only contesting the powers that be but also directly creating the change they desire. As we can see, the contentious politics of data places these activists in a unique structural and cultural position, intrinsically intertwined with digital data and technology. These factors make them particularly fitting for understanding the changing nature of activism as it is mediated by data.

Additionally, these two cases were chosen to capture a range of dynamics in data activism. While they share some similarities, they also differ in important ways. Namely, DR is a mainstream movement made up of formally structured organizations which typically act within the boundaries of the law and focus on policy, combining online digital activism with offline activities in legal institutions. In contrast, ANON is known for being anarchic and leaderless, with participants tending to favor more confrontational and disruptive digital tactics. I examine both cases to determine how, if at all, they differ from what the established literature would lead us to expect. Below, I provide brief backgrounds for the two cases.

### *The digital rights movement (DR)*

The Digital Rights movement was born from the countercultural and cyber libertarian values of early Internet culture (Levy, 1984). It first came into conflict with the law in the 1980s over Free/Open-Source Software (F/OSS), or software that is free of any private ownership and open to anyone who wishes to use it. It was intended as an alternative to closed, proprietary software systems produced by corporations such as Microsoft and Apple. This led to the creation of the Free Software Foundation (FSF) in 1985 and the Electronic Frontier Foundation (EFF) in 1990. In 1999, a new and more visible level of contentious politics erupted over an open-source Linux program called Decrypt Content Scrambling System (DeCSS). According to Coleman (2012), hackers and coders had acquired new legal expertise through their experience with contested F/OSS projects which they used to mobilize and win a major court case, propelling the movement forward into greater legitimacy.

Today, datafication and dataveillance have also now become leading issues of digital rights activism. The movement has grown to consist of numerous organizations, of which the EFF is one of the most central and widely recognized. Originally a small non-profit legal defense group, it has become an influential civil society organization of lawyers, lobbyists, technical advisors, coders, and activists. They engage in lawsuits, lobbying, protest campaigns (including petitions, boycotts, website blackouts, and tweet storms), and technological innovation (such as privacy software development) with the goal of extending civil and human rights into the realm of digital data and technology. DR organizations have mobilized largescale grassroots support, such as the campaigns against the Protect IP Act (PIPA) and the Stop Online Piracy Act (SOPA) in 2012 (Daskal, 2018). DR especially gained greater awareness and emotional solidarity after the death of hacker and activist Aaron Swartz in 2013, who became memorialized as a hero and martyr for the movement (Martini, 2018).

### *Anonymous (ANON)*

Beginning around 2006, another group emerged online from an unexpected corner of Internet subculture. From the imageboard website 4chan, the concept of Anonymous as a collective identity evolved like an Internet meme (Beyer, 2014; Coleman, 2014). Initially, 'Anons' began using this collective identity to go on virtual 'raids' of other online websites and communities for fun, or 'lulz.' However, in 2009, the collective underwent a transformation during its first major activist campaign, a protest of the Church of Scientology known as 'Project Chanology.' This campaign demonstrated the potential of Anonymous as a collective political actor and drew in large numbers of new participants focused on more activist-oriented causes (Coleman, 2014). Anonymous protest campaigns, which arguably peaked in activity from 2009 to 2012, are primarily known for 'hacktivism,' or the use of computer hacking tools and techniques for political activist goals (Samuel, 2004; Tilly & Tarrow, 2007). While Anons are heterogeneous in terms of ideology (Fuchs, 2013; Goode, 2015), many of their largest collective action campaigns during this period reflected issues such as freedom of speech, freedom from censorship, and the freedom of information, especially those coming through prominent subgroups like AnonOps (Coleman, 2014).

Scholars have debated whether Anonymous should be thought of as a new type of social movement, a collective collaboration, or an activist ensemble (Coleman, 2014, 2017; Deseriis, 2013; McDonald, 2015; Uitermark, 2017). In practice, it is a complex system of subgroups fragmented into 'cabals nested within cabals' (Coleman, 2014, p. 114). These subgroups at their core typically consist of small cadres of computer hackers surrounded by a much larger group of what Coleman calls 'geeks' who possess a broad array of digital media literacies. Coleman (2014) contends that while participants may come and go over time, and key nodes may rise and fall, the cultural codes and 'overarching ideals remain in play' (p. 48). However, others have argued that the Anonymous subgroups are far more disparate (Uitermark, 2017) and that in recent years the Anonymous 'brand' has increasingly been co-opted by a highly diverse array of groups for many different, and at times conflicting, issues (Beraldo, 2020).

## The perception and framing of opportunity and threat

In this study, I use the aforementioned cases to examine the dynamics of *political opportunities* and *threats*, two central concepts from theoretical work on social movements. Both describe changes in the broader environment in which movements exist that can increase the likelihood of mobilization. Political opportunities are openings or vulnerabilities in the political structure which can lower the costs of collective action, including increasing access, shifting alignments, divided elites, and influential allies

(McAdam et al., 2001; McAdam, 1982; Meyer & Minkoff, 2004; Tarrow, 1998). Further theorizing suggests differences between *stable* political opportunity structures, such as aspects of nation-states and formal institutions which only change very slowly and with great effort, and *volatile* political opportunities, which shift more rapidly within protest cycles (Della Porta, 1995; Gamson & Meyer, 1996).

If opportunity is an opening which *lowers the cost of action*, then threat is a force which *increases the cost of inaction.* (Einwohner & Maher, 2011; Maher, 2010; Van Dyke & Soule, 2002). Goldstone and Tilly (2001) define threat as, 'the cost that a social group will incur from protest, or that it expects to suffer if it does not take action' (p. 138). In this manner, opportunity can be understood as a 'pull' while threat can act as a 'push' to mobilize (Einwohner & Maher, 2011, p. 129). While opportunity and threat may both arise across waves of protest (Almeida, 2008), threat is not merely the inverse of opportunity; they are conceptually distinct (Tarrow, 1998). There is still much debate in the literature regarding under what conditions or contexts each might be more influential (Johnson & Frickel, 2011; Shriver et al., 2015).

However, this understanding of opportunities has been criticized as overly focused on structure and resources to the neglect of culture and agency (Goodwin & Jasper, 1999). Some have argued that opportunities and threats must not merely exist but also be perceived and interpreted (Meyer & Minkoff, 2004; Tarrow, 1998). These interpretations are then strategically framed and communicated to potential participants to mobilize collective action. A *frame* is a 'schemata of interpretation' that enables individuals 'to locate, perceive, identify, and label' an issue, problem, or event (Goffman, 1974, p. 21). Frames function like a 'story line' for people to create meaning and thus guide their actions (Gamson & Wolfsfeld, 1993). According to Johnston (2002), frames are both fixed cognitive structures and emergent cognitive processes, existing as both a noun 'frame' and a verb 'framing.' Framing as a verb denotes 'meaning work' or an 'active, processual phenomenon that implies agency and contention at the level of reality construction' (Benford & Snow, 2000, pp. 613–614). While framing can take place at the individual level, social movements are known to strategically craft *collective action frames* (Snow & Benford, 1992) as a micro-mobilization process to, 'frame or assign meaning to and interpret relevant events and conditions in ways that are intended to mobilize potential adherents and constituents, to garner bystander support and to demobilize antagonists' (Snow & Benford, 1988, p. 198).

Collective action frames are generated through discursive and at times contested processes. Working to synthesize the macro-level influence of opportunity structures with the micro-mobilization view of frames, Koopmans and Olzak (2004) argue that cultural elements in a movement's broader environment can facilitate or constrain which collective action frames will be most successful. These discursive opportunity structures (McCammon, 2013) shape which frames are successful based on how they are diffused through mass media and resonate with external audiences and potential followers in the public sphere. Furthermore, the concept of the mediation opportunity structure goes further by combining media, discursive, and networked opportunities (Cammaerts, 2012). It suggests that activists must contend with the dominant frames of often biased mainstream media and political elites, create their own discursive self-representations which resonate with the public, and utilize the affordances and constraints of networked technologies for organizing, counter-framing, and engaging in direct action.

Combining these structural and cultural perspectives, I ask how political opportunities and threats are perceived and framed in the context of contentious data politics, with digital data either as the repertoire or the stake of activism. First, considering data as repertoire, the practices of data activists potentially enable them to highly leverage digital affordances, from so-called 'clicktivism' (Morozov, 2011) to direct hacktivism, thereby lowering the costs of action. Therefore, are frames of political opportunities, which also lower the costs of action, still salient for already low-cost data activism? Second, regarding data as stakes, research has shown that top-down technological changes, such as datafication and algorithmic governance, tend to be dominantly framed as 'progress' by the powers that be in the public sphere. In response, activists often stress the 'potential hazardous consequences' of technology (Daskal, 2018, p. 7) to build their 'counter-imaginaries' (Kazansky & Milan, 2021) of digital rights and freedom. Thus, do data activist rely mostly on frames of threat to mobilize collective action? The methodological steps I took to answer these questions are described in the following section.

## Data and methods

For this study, I utilized qualitative thematic coding and analysis (Berg & Lune, 2012) of four years of textual archival documents from each of my cases. NVivo software was used to organize the data which I manually coded (MacMillan & Koenig, 2004). My coding process was based on a modified grounded theory approach, combining both deductive and inductive coding in multiple iterative passes (Bringer et al., 2006; Strauss, 1987). I first categorized the main topics being discussed (e.g. privacy, free speech), then coded for how they were framed, or assigned meaning and communicated in ways that are intended to garner public support and mobilize potential adherents (Snow & Benford, 1988, 1992). Specifically, I coded whether they were framed as changes that would facilitate or constrain a movement's ability to advance its rights (political opportunities) or that would take away extant rights unless stopped (threats). Under opportunities, I also created child codes to differentiate between direct, immediate effects (volatile) and broader, structural effects (stable). In some cases, portions of text were coded in multiple categories.

The data consisted of public documents from a primary group within each case. These were documents which were issued on behalf of the group to an audience of their own followers and the general public. For the Digital Rights movement, I analyzed newsletters from the Electronic Frontier Foundation (EFF) called the 'EFFector' which is a pun on a computer science term meaning, 'a device for producing a desired change.' All issues were publicly available on the EFF's website (EFF, 2021). For Anonymous, I analyzed the press statements issued by one of the largest and most active public outlets, AnonNews. These statements are perhaps best known in video form with a person wearing a Guy Fawkes mask reading through a digitally altered voice. According to Coleman (2014), these statements are written through a deliberative process involving numerous Anon authors. AnonNews maintained a website up until 2012 which kept a public textual archive of every press statement issued on behalf of Anonymous starting with 2009's Project Chanology (AnonNews, 2012). AnonNews still maintains a Twitter account today and has been associated with other prominent subgroups such as AnonOps, one of the most prolific nodes in Anonymous (Coleman, 2014).

Although these two sources of data are not identical, they both offered a view into how activists understand and frame specific issues and events. I chose to examine documents which were issued from 2009 to 2012 because it represents a high point of activity for both cases. DR's numerous campaigns against surveillance and intellectual property restrictions, and mass protests of the international trade deal known as ACTA occurred during this time (Coleman, 2012; Postigo, 2012). In addition, some of the largest ANON campaigns also occurred therein, such as Operation Payback as well as Operations Tunisia and Egypt during the Arab Spring uprisings (Coleman, 2014).

During this period, the EFF website contained 135 issues of its 'EFFector' newsletter. Each contained short summaries of recent press statements and news stories, which linked to longer in-depth articles. For this same period, the AnonNews archive contained approximately 200[2] unique press statements. Each statement was typically written in the style of a letter, addressed to a specific audience (e.g. 'People of Tunisia', 'Fellow Anons', or 'Members of NATO'). Unlike the EFFector newsletters, which were on average three to four pages long and covered several topics in each issue, the Anonymous press statements were shorter (mostly one to three pages) and addressed only a single issue or 'operation' at a time. The structure and tone of each corpus in many ways strongly reflected the attitudes and ideas of each group. For the purposes of this study, these documents were well-suited data and strong enough to reveal patterns in how movement activists framed opportunities and threats.

To conduct a detailed qualitative analysis, I chose to draw representative samples from each corpus. If we treat individual documents as the unit of analysis and consider all EFFector newsletters and AnonNews statements as the respective populations, then the 135 Effector newsletters and 200 AnonNews statements between 2009 and 2012 are my sampling frame. I randomly sampled from each, using a random number generator. Based on an initial analysis, the sampling continued until I determined that I had reached a point of sufficient data saturation (Glaser & Strauss, 1967; Strauss, 1987). I began with the AnonNews statements. After eliminating duplicates, I ended up with a sample of 50 out of 200 unique statements (25%).[3] From the EFFector newsletters, I sampled a total of 41 out of 135 (30%). Starting with DR, key examples of passages I coded are presented in the following section to exemplify the prominent themes that emerged.[4]

## Findings

### *Opportunity frames in the digital rights movement*

According to NVivo, I coded a total of 18.44% of the EFF texts for frames of *Opportunity* and 34.81% for *Threat*. Although only about half as frequent as threat, political opportunity frames still played an important role for the EFF. Often, these discussed how technological advances have created broad openings for all types of activists to communicate and organize to advance their causes. This theme of Internet technology as a *stable* political opportunity recognized by the authors was evident in passages such as, 'Websites like YouTube have ushered in a new era of creativity and free speech on the Internet' (lines 877–8) and 'The advent of advanced communications technologies offers an opportunity to improve responsiveness and accessibility . . . ' (lines 2319–20).

The texts also revealed that, in addition to many online forms of activism, they engaged a great deal with lawsuits and policy formation which took place in courtrooms and other legal institutions. Most of the references to *volatile* political opportunities in the EFF texts pertained to the legal system, such as favorable rulings in court cases or changes to policies which created new openings for the EFF and its allies to push for increased digital rights. For example, one passage about a speech in January 2010 stated, "Secretary Clinton's speech on Internet Freedom was an important step in bringing online free expression and privacy to the forefront of the United States' foreign policy agenda" (lines 1361–2). Here we can see how the support of a political elite helped advance digital rights in the federal policy arena, a well-known form of political opportunity (Tarrow, 1998).

Another passage described how the EFF was able to use the legal system to help pass a bill to protect online privacy rights: 'In another big step towards updating reader privacy for the digital age, the California State Senate Judiciary Committee passed through SB 602, the Reader Privacy Act, after hearing testimony from EFF Legal Director Cindy Cohn' (lines 2550–2). Such victories created new openings for DR activists. But changes in policy did not necessarily have to be victories to be seen as opportunities. Sometimes, perceived opportunities and threats were mixed in the texts. As the next passage shows, activists framed a Congressional vote supporting a threatening bill as an opportunity to engage in more 'meaningful' protest and reform: 'Congress voted at the end of February to extend the expiring provisions of the USA PATRIOT Act, but only for three months – giving liberty-loving citizens a new opportunity to continue to protest this dangerous law and demand the passage of meaningful PATRIOT reform before the end of May' (lines 2239–41). Whether through support from elite allies, testimony from legal experts, or protest activity from its grassroots members, the open legal system was framed by the EFF as a focal point of opportunity to create social and political change.

### *Threat frames in the digital rights movement*

While opportunities are important for the EFF, nearly twice as frequent were frames about threats that could take away digital rights unless something is done to stop them. References to the threats posed by 'copyright trolls' made up the most common recurring theme. For example, 'Our movie industry has created some memorable monsters on screen. But Hollywood and the major music labels also helped create a very real kind of monster: copyright trolls who coerce settlements from Internet subscribers using intimidation and our out-of-whack copyright laws' (lines 3510–12). Another recurring topic was ACTA (Anti-Counterfeiting Trade Agreement of 2011), which was framed as a policy that, in effect, would act as a form of censorship threatening people's freedom and privacy unless stopped: "ACTA raises serious concerns about citizens' civil liberties and privacy rights. The contents and text of ACTA remain secret, but a document leaked to the public last year shows that ACTA could include stronger criminal measures, increased customs border search powers, and requirements for Internet service providers to cooperate with copyright holders" (lines 15–21).

Additional threats were mostly derived from government and corporate surveillance, data tracking, and oppression of Internet users, which were typically framed as threats to privacy, liberty, and/or freedom (of speech, expression, creativity, sharing). For instance, certain legislation was framed as threatening digital rights by expanding government authority to openly monitor the online activities of its citizens with little oversight: 'One of the provisions of the Media and Wiretapping Bill currently being discussed by the Italian Parliament threatens free expression and innovation in Italy' (lines 2980–1). A similar passage states, 'Australians are fending off threats to their right to privacy from all directions' (lines 3836) due to expanding government online surveillance powers and the Cybercrime Legislation Amendment Bill 2011. Threats of arrest and even bodily harm towards those critical of government in authoritarian states also emerged as a common theme: 'The intimidation and persecution of bloggers and online journalists is a grave threat to free expression in many countries these days. The effects are often far-reaching as bloggers are scared into silence. Bloggers from the UAE, Egypt, and Syria have all been threatened with prosecution this week' (lines 3029–32).

From this analysis, we can see that the EFF was a highly engaged DR organization that worked to influence digital policy both in the legal arena and through collective action. The EFF combined both online activism and offline legal actions. While they discussed some stable political opportunities regarding technology, their volatile opportunity frames primarily focused on offline legal activities. Nonetheless, the frequency of opportunity frames was still only half that of threat.

### Opportunity frames in anonymous

I coded 7.3% of the AnonNews corpus for *Opportunity*, and 23.62% for *Threat*. Here, threat frames appeared with more than three times greater frequency than opportunity. Frames of opportunity were rare overall and focused entirely on *stable* political opportunities. These were typically framed as advancements in Internet technology which created a broad historical opening for activism and social change. For example, 'Since its inception, the internet has provided new ways for people all over the world to exercise the rights of free speech, freedom of the press and freedom of assembly' (lines 488–9). Another described how Internet technology connects people around the world by overcoming traditional barriers to organizing: 'With the advent of the internet . . . society – and the way it is organized – has undergone a dramatic change, the consequences of which are not yet completely visible. With this technology has come the first true opportunity to access, share, discuss, and produce information anywhere and anytime, transcending previous boundaries of locality and temporality, as well as certain socio-economic limitations' (lines 1438–42).

Other stable opportunity frames came in the form of perceived support from the media. For example, in a statement in early 2011, Anons directly addressed members of the media. 'You [the media] were intrigued by us when we took on the financial sector by attacking Visa, Mastercard, and Paypal as part of Operation Payback. Many of you became enamoured with us during Operation Tunisia and applauded our efforts. Some of you are now paying close attention to Operation Algeria. We know that we have gained support from even a few of you' (lines 522–5). Attention and support from the media can be a significant opportunity for activists to spread awareness and sway public opinion on issues. Another statement similarly addressed Al Jazeera for their dedicated reporting on the Arab Spring protests in Egypt, stating, 'Thank you, Al Jazeera, for your outstanding coverage in the streets of Egypt. Your constant reporting and unbiased journalism has helped unite the revolutionaries, and the world' (lines 1332–3).

Much less frequent were references to *volatile* political opportunities within specific campaigns. These included divisions among elites or state forces opposing protesters, another well know form of political opportunity. For example, this passage from a statement about OpTunisia said, 'Police officers were being held up by a crowd and kissed – some police had started changing sides' (line 825). Another passage in a different statement about the Tunisian protests stated, 'Anonymous would like to applaud the head of the Tunisian military – Rashid Ammar – for risking his personal security by refusing to fire on Tunisian demonstrators' (lines 1716–19). As we can see, police and military forces switching sides to support protesters was a major opportunity during the protests. However, both quotes referred to opportunities for the protesters in Tunisia rather than for Anons themselves. The corpus did not contain any frames of volatile opportunities directly impacting Anons.

### *Threat frames in anonymous*

In contrast, frames of threat appeared far more frequently in the texts. Threats comprised nearly a quarter of the entire corpus and were primarily framed as actions which deprive people freedom of speech. For example, as one statement clearly put it, 'Free Speech is the most fundamental of your fundamental rights. Once you lose it, you won't have the means to complain when politicians come back and destroy your wealth, your health, your soul, your ideals, your hopes, your projects, your whole life. Free Speech is the freedom that protects all freedoms' (lines 435–8). This theme of 'freedom of speech' appeared in nearly every statement, often equated with the freedom of the Internet itself. For example, 'The future of the internet hangs in the balance' (line 91), 'The trap of international censorship is desperately trying to swallow the Internet' (lines 459–60), and ' . . . most human beings live under censorship – and you're next on the list, my dear Anon friend' (lines 478–9). This next passage shows exactly how Anons perceived online censorship as a threat, 'Our aim is to force them to abandon censorship on the Internet. Censorship is stupid, harmful, dangerous, expansive, unuseful, pathetic. It stinks like hell. We treat censorship as a deadly enemy, a mental illness, as symbol of totalitarianism, a practice of power that must die painfully' (lines 307–10).

In other cases, threats were seen as affecting groups on whose behalf Anons were engaging. Anons clearly acknowledged their supporting role as outsiders in many cases. For example, in reference to a proposed law in Spain known as 'Ley Sinde,' Spanish hacktivists implored the Anonymous collective to help them by describing how the law will 'not only restrict our fundamental rights (freedom of speech), but will also be committing an evident act of censorship' (lines 581–5). Last, this passage described how 'Free-Thinking Citizens of the World' needed to come together to stop the Algerian government from oppressing its people. 'The Algerian government hides the truth from its citizens and denies them free expression and access to information . . . Restrictions on the freedom of speech and denial of information to your citizens are in direct violation of the Universal Declaration of Human Rights, and cannot be tolerated' (lines 741–2).

The perceived cost of inaction (i.e. threat) was emphasized strongly and clearly throughout the texts. For instance, 'ANONYMOUS therefore believes the Tunisian attempts at censorship are doomed to failure if only we, ANONYMOUS, the people, take up our individual responsibilities. For if only we decide to make it so – it will be done' (lines 350–2). Here, they were saying explicitly that the censorship practices of the state would fail only if Anons acted to stop them. This was a call to action, not based on an opportunity, but on a perceived threat which implied a clear cost of inaction (i.e. censorship). Yet, there was no mention of volatile opportunities for Anons, in this statement or any other, perhaps because the cost of action was already perceived as being so low by Anons that volatile political opportunities served no useful function for them at all. Instead, the texts emphasized confidence and efficacy to express Anonymous' capacities via technological affordances. For instance, the following quote reflects the type of wording used throughout all Anon statements, 'We are not quiet. We are loud and when the people roar it quakes the foundations with which these Governments stand. We are Citizens and if we choose: We are Anonymous' (lines 252–5).

## Discussion and conclusion

This study has explored how data mediate the perception and framing of political opportunities and threats by examining two cases of contentious data politics. My findings show that, first, threat plays a much more prominent role than opportunity in the collective action frames of both cases. While much influential social movement scholarship has emphasized the importance of political opportunity for mobilization (McAdam et al., 2001; McAdam, 1982; Tarrow, 1998), these findings call such emphasis into question. The relatively more recent and less researched concept of threat has been found to mobilize action primarily in specific contexts, including under circumstances of authoritarianism and oppression (Almeida, 2008; Einwohner & Maher, 2011; Goldstone & Tilly, 2001; Maher, 2010), in far-right and fascist groups (McVeigh, 2009; Van Dyke & Soule, 2002; Wright, 2007), and in environmental activism (Johnson & Frickel, 2011; Shriver et al., 2015). This study shows empirically that frames of threat are also more prominent than opportunity in the context of 'Internet freedom' contentious data politics.

Second, while frames of opportunity were less common than threat in both cases, there was also an observed difference between the two. The EFF had both stable and volatile opportunity frames, the latter of which were primarily related to offline activities in legal institutions, such as lobbying or litigation. The Anons, whose hacktivism primarily took place online, had a only small proportion of stable opportunity frames, yet no frames of volatile opportunities directly affecting themselves. When it came to mobilizing online actions ('data as repertoire'), both groups primarily used frames of threats. Even when discussing tactics, they correspondingly emphasized more 'reactive' data activism, characterized by adversarial and defensive forms of resistance such as encryption and anonymity practices (Milan & van der Velden, 2016), especially in the Anonymous texts.

These findings suggest the possibility that not only are perceived threats more important in contentious data politics, but that political opportunities also appear to have lower salience for online activism. After all, an opportunity not seen is an opportunity that, for all intents and purposes, does not exist (Meyer & Minkoff, 2004). One possible explanation is that, in face-to-face actions such as a protest rally or lawsuit, the associated costs and risks can be very high, thus volatile political opportunities which can lower the immediate costs are very important. In contrast, activism that takes place within the relative detachment and anonymity of digital space, such as hacking a website, can leverage technological affordances to lower such costs. Thus, volatile political opportunities may be perceived as less important as a 'pull' towards action compared to the 'push' of threats.

But there are cultural factors present as well. For example, the networked opportunity perspective can help us see not just structural affordances but also the practices by which some activists are able to leverage affordances to a higher degree, while media and discursive opportunities also enable us to consider how frames are constructed to counter dominant frames in the media and society, potentially accounting for why threat frames are so prevalent in both cases. Scholars of contemporary social movements should thus rethink the centrality of political opportunity structures that characterizes much of the literature. Opportunity and threat need to be better connected into a unified theoretical framework which also integrates both structural and cultural factors.

This study shows how data can mediate collective action dynamics as both the repertoire and stakes. But such influences are likely to become more widespread in the future as a greater variety of mobilizations begin to adopt these same data practices and issues. In some ways, these data activists fighting for what they see as the 'Freedom of the Internet' have created the model for digital activism which others will have to follow as they migrate more and more into digital space. It was early Digital Rights activists who first fought to establish free speech protections online many years before social media were invented, and it was Anonymous hacktivists who helped popularize the tactics of digital sit-ins which are now being deployed by TikTok teens, K-Pop stans, and Black Lives Matter activists, among others. Whether they are aware or not, this data activism set the stage for all the digitally enabled activism that will come afterwards.

This study builds on the heuristic concept of contentious data politics (Beraldo & Milan, 2019) and contributes to the literature by showing empirically how data can mediate the structural and cultural factors that determine how opportunities and threats are perceived and framed in contemporary activism. The approach of this study was to compare these cases with existing theories of collective action dynamics which were originally formed with more traditional social movements in mind. Future studies of this nature could be carried out on additional types of online and offline movements to support whether opportunity frames are indeed as important across most types of movements as the extant body of literature appears to suggest. Additional studies could also examine other collective action dynamics to discover further changing or evolving processes.

This study is not without its limitations. For one, the sample size for the texts of AnonNews ended up being slightly smaller than that of the EFF. While a larger sample of Anon texts could be more revealing, it was not possible since the original archive from which the texts were drawn no longer exists. However, given that nothing ever truly disappears from the Internet, it may be possible to track down these or additional Anonymous press statements through other sources, possibly using a targeted sampling technique. Moreover, because the sample randomly draws from four years, it is possible that more active periods may be overrepresented. For example, some Anonymous operations, such as during the Arab Spring, may have generated more press statements than others. Last, most studies of these movements have focused on ethnographic approaches or social media analyses; examining their archival texts is a novel contribution of this study. Yet, it is also a limitation. Additional studies using different data, mixed methods, or alternate periods of time could provide further insights into these cases.

## Notes

1. This is a play-on-words referring to Scott's (1985) concept of "weapons of the weak" which described how poor rural peasants converted their farming tools into weapons to rebel against much more powerful authorities. Coleman uses a similar line of reasoning to explain how the hacker and coder community, an otherwise apolitical and privileged group, has become politically influential in recent years.
2. The full archive at the time of my sampling contained 412 statements; however, approximately half of them turned out to be duplicates. For example, a single statement would often appear multiple times translated into different languages.
3. My initial sample of AnonNews statements was taken in 2012. Soon afterwards, the archive was taken permanently offline. For this reason, I was unable to subsequently expand upon the existing sample.
4. The sampled texts were combined into a single text document for each corpus. The quotations presented in the Findings section are cited by the line number in which they appeared in their respective corpus.

## Acknowledgement

Special thanks to Rachel L. Einwohner, S. Laurel Weldon, Bert Useem, Stuart A. Wright, Andrew Raridon, Soon Seok Park, and Gulcin Con Wright.

## Disclosure statement

No potential conflict of interest was reported by the author(s).

## Funding

This project was funded in part by the Purdue University Walter Hirsch Graduate Student Dissertation Research Award and the Purdue Research Foundation Dissertation Grant.

## ORCID

Jared M. Wright http://orcid.org/0000-0002-8289-2000

## References


Almeida, P. D. (2008). *Waves of protest: Popular struggle in El Salvador, 1925-2005*. University of Minnesota Press.

AnonNews. (2012). http://www.anonnews.org.

Benford, R. D., & Snow, D. A. (2000). Framing processes and social movements: An overview and assessment. *Annual Review of Sociology, 26*(1), 611–639. https://doi.org/10.1146/annurev.soc.26.1.611

Bennett, W. L., & Segerberg, A. (2013). *The logic of connective action: Digital media and the personalization of contentious politics*. Cambridge University Press.

Beraldo, D. (2020). Movements as multiplicities and contentious branding: Lessons from the digital exploration of #occupy and #anonymous. *Information, Communication & Society 25*(8), 1098–1114. https://doi.org/10.1080/1369118X.2020.1847164.

Beraldo, D., & Milan, S. (2019). From data politics to the contentious politics of data. *Big Data & Society, 6*(2). https://doi.org/10.1177/2053951719885967

Berg, B. L., Lune, H. 2012 Qualitative research methods for the social sciences 8th (Boston, MA: Pearson Education, Inc.)9780205234967

Beyer, J. L. (2014). *Expect us: Online communities and political mobilization*. Oxford University Press.

Bringer, J. D., Johnston, L. H., & Brackenridge, C. H. (2006). Using computer-assisted qualitative data analysis software to develop a grounded theory project. *Field Methods, 18*(3), 245–266. https://doi.org/10.1177/1525822X06287602

Cammaerts, B. (2012). Protest logics and the mediation opportunity structure. *European Journal of Communication, 27*(2), 117–134. https://doi.org/10.1177/0267323112441007

Castells, M. (2015). *Networks of outrage and hope: Social movements in the internet age* (2nd ed.). John Wiley & Sons.

Coleman, G. (2012). *Coding freedom: The ethics and aesthetics of hacking*. Princeton University Press.

Coleman, G. (2014). *Hacker, hoaxer, whistleblower, spy: The many faces of anonymous*. Verso books.

Coleman, G. (2017). From internet farming to weapons of the geek. *Current Anthropology, 58* (S15), S91–S102. https://doi.org/10.1086/688697

Daskal, E. (2018). Let's be careful out there . . . : How digital rights advocates educate citizens in the digital age. *Information, Communication & Society, 21*(2), 241–256. https://doi.org/10.1080/1369118X.2016.1271903

Della Porta, D. (1995). *Social movements, political violence, and the state: A comparative analysis of Italy and Germany*. Cambridge University Press.

DeNardis, L. (2012). Hidden levers of internet control: An infrastructure-based theory of internet governance. *Information, Communication & Society, 15*(5), 720–738. https://doi.org/10.1080/1369118X.2012.659199

Dencik, L., Hintz, A., & Cable, J. (2016). Towards data justice? The ambiguity of anti-surveillance resistance in political activism. *Big Data & Society, 4*(2), 1–12.

Deseriis, M. (2013). Is anonymous a new form of Luddism? A comparative analysis of industrial machine breaking, computer hacking, and related rhetorical strategies. *Radical History Review, 2013*(117), 33–48. https://doi.org/10.1215/01636545-2210437

Diani, M., & Bison, I. (2008). Organizations, coalitions, and movements. *Theory and Society, 33*(3/4), 281–309. https://doi.org/10.1023/B:RYSO.0000038610.00045.07

Earl, J., & Kimport, K. (2009). Movement societies and digital protest: Fan activism and other nonpolitical protest online. *Sociological Theory, 27*(3), 220–243. https://doi.org/10.1111/j.1467-9558.2009.01346.x

Earl, J., & Kimport, K. (2011). *Digitally enabled social change: Activism in the internet age*. MIT Press.

EFF. (2021). *Effector*. Electronic Frontier Foundation. https://www.eff.org/effector.

Einwohner, R. L., & Maher, T. (2011). Threat assessment and collective-action emergence: Death-camp and ghetto resistance during the Holocaust. *Mobilization: An International Quarterly, 16*(2), 127–146. https://doi.org/10.17813/maiq.16.2.j263166u14286024

Fish, A., & Follis, L. (2016). Gagged and doxed: Hacktivism's self-incrimination complex. *International Journal of Communication, 10*, 20 https://ijoc.org/index.php/ijoc/article/view/5386.

Flesher Fominaya, C., & Gillan, K. (2017). Navigating the technology-media-movements complex. *Social Movement Studies, 16*(4), 383–402. https://doi.org/10.1080/14742837.2017.1338943

Fuchs, C. (2013). The anonymous movement in the context of liberalism and socialism. *Interface: A Journal for and About Social Movements, 5*(2), 345–376.

Gamson, W. A., & Meyer, D. S. (1996). Framing political opportunity. In D. McAdam, J. D. McCarthy, & M. N. Zald (Eds.), *Comparative perspectives on social movements: Political opportunities, mobilizing structures, and cultural framings* (pp. 275–290). Cambridge University Press.

Gamson, W. A., & Wolfsfeld, G. (1993). Movements and media as interacting systems. *The Annals of the American Academy of Political and Social Science, 528*(1), 114–125. https://doi.org/10.1177/0002716293528001009

Gerbaudo, P. (2012). *Tweets and the streets: Social media and contemporary activism.* Pluto Press.

Glaser, B. G., & Strauss, A. L. (1967). *Discovery of grounded theory: Strategies for qualitative research.* Macmillan.

Goffman, E. (1974). *Frame analysis: An essay on the organization of experience.* Harvard University Press.

Goldstone, J. A., & Tilly, C. (2001). Threat (and opportunity): Popular action and state response in the dynamics of contentious action. In R. Aminzade, D. McAdam, W. H. Sewell, J. A. Goldstone, S. Tarrow, E. J. Perry, & C. Tilly (Eds.), *Silence and voice in the study of contentious politics* (Cambridge University Press) (pp. 179–194).

Goode, L. (2015). Anonymous and the political ethos of hacktivism. *Popular Communication, 13*(1), 74–86. https://doi.org/10.1080/15405702.2014.978000

Goodwin, J., & Jasper, J. M. (1999). Caught in a winding, snarling vine: The structural bias of political process theory. *Sociological Forum, 14*(1), 27–54. https://doi.org/10.1023/A:1021684610881

Greenwald, G. (2014). *No place to hide: Edward Snowden, the NSA, and the U.S. surveillance state.* Metropolitan Books/Henry Holt.

Guidry, J., Kennedy, M. D., & Zald, M. N. (Eds.). (2000). *Globalizations and social movements: Culture, power, and the transnational public sphere.* University of Michigan Press.

Habermas, J. (1989). *The structural transformation of the public sphere: An inquiry into a category of bourgeois society* T. Burger (Ed.). MIT Press. Original work published 1962.

Highfield, T. (2016). *Social media and everyday politics.* John Wiley & Sons.

Hu, M. (2017). From the national surveillance state to the cyber surveillance state. *Annual Review of Law and Social Science, 13*(1), 161–180. https://doi.org/10.1146/annurev-lawsocsci-110316-113701

Johnson Andrews, S. (2019). *The cultural production of intellectual property rights: Law, labor, and the persistence of primitive accumulation.* Temple University Press.

Johnson, E. W., & Frickel, S. (2011). Ecological threat and the founding of US national environmental movement organizations, 1962–1998. *Social Problems, 58*(3), 305–329. https://doi.org/10.1525/sp.2011.58.3.305

Johnston, H. (2002). Verification and proof in frame and discourse analysis. In B. Klandermans & S. Staggenborg (Eds.), *Methods of social movement research* (pp. 62–91). University of Minnesota Press.

Kazansky, B., & Milan, S. (2021). Bodies not templates: Contesting mainstream algorithmic imaginaries. *New Media & Society, 23*(2), 363–381. https://doi.org/10.1177/1461444820929316

Koopmans, R., & Olzak, S. (2004). Discursive opportunities and the evolution of right-wing violence in Germany. *The American Journal of Sociology, 110*(1), 198–230. https://doi.org/10.1086/386271

Lessig, L. (2006). *CODEv2*. Basic Books.

Levy, S. (1984). *Hackers: Heroes of the computer revolution*. Anchor Press/Doubleday.

MacMillan, K., & Koenig, T. (2004). The wow factor: Preconceptions and expectations for data analysis software in qualitative research. *Social Science Computer Review*, *22*(2), 179–186. https://doi.org/10.1177/0894439303262625

Maher, T. (2010). Threat, resistance, and mobilization: The cases of Auschwitz, Sobibór, and Treblinka. *American Sociological Review*, *75*(2), 252–272. https://doi.org/10.1177/0003122410365305

Martini, M. (2018). Mourning for a hacktivist: Grieving the death of Aaron Swartz on a digital memorial. *Media, Culture & Society*, *40*(2), 228–245. https://doi.org/10.1177/0163443717718254

McAdam, D. (1982). *Political process and the development of black insurgency, 1930-1970*. University of Chicago Press.

McAdam, D., Tarrow, S., & Tilly, C. (2001). *Dynamics of contention*. Cambridge University Press.

McCammon, H. J. (2013). Discursive opportunity structure Snow, D. A., Della Porta, D., Klandermans, B., McAdam, D. In *The Wiley-Blackwell encyclopedia of social and political movements*. John Wiley & Sons.

McDonald, K. (2015). From Indymedia to anonymous: Rethinking action and identity in digital cultures. *Information, Communication & Society*, *18*(8), 968–982. https://doi.org/10.1080/1369118X.2015.1039561

McVeigh, R. (2009). *The rise of the Ku Klux Klan: Right-wing movements and national politics*. University of Minnesota Press.

Meyer, D. S., & Minkoff, D. C. (2004). Conceptualizing political opportunity. *Social Forces*, *82*(4), 1457–1492. https://doi.org/10.1353/sof.2004.0082

Meyer, D. S., & Tarrow, S. (Eds.). (1998). *The social movement society: Contentious politics for a new century*. Rowman & Littlefield.

Milan, S. (2015a). From social movements to cloud protesting: The evolution of collective identity. *Information, Communication & Society*, *18*(8), 887–900. https://doi.org/10.1080/1369118X.2015.1043135

Milan, S. (2015b). When algorithms shape collective action: Social media and the dynamics of cloud protesting. *Social Media + Society*, *1*(2). https://doi.org/10.1177/2056305115622481

Milan, S., & Ten Oever, N. (2017). Coding and encoding rights in internet infrastructure. *Internet Policy Review*, *6*(1), 1–17. https://doi.org/10.14763/2017.1.442

Milan, S., & van der Velden, L. (2016). The alternative epistemologies of data activism. *Digital Culture & Society*, *2*(2), 57–74. https://doi.org/10.14361/dcs-2016-0205

Morozov, E. (2011). *The net delusion: How not to liberate the world*. Penguin.

Park, S. S., & Einwohner, R. L. (2019). Becoming a movement society? Patterns in the public acceptance of protest, 1985–2006. *Sociological Focus*, *52*(3), 186–200. https://doi.org/10.1080/00380237.2019.1624233

Postigo, H. (2012). *The digital rights movement: The role of technology in subverting digital copyright*. MIT Press.

Samuel, A. W. (2004). *Hackivism and the future of political participation*. (Doctoral dissertation, Harvard University).

Schmidt, E., & Cohen, J. (2014). *The new digital age: Transforming nations, businesses, and our lives*. Random House.

Schradie, J. (2019). *The revolution that wasn't: How digital activism favors conservatives*. Harvard University Press.

Scott, J. 1985.*Weapons of the weak: Everyday forms of peasant resistance*. Yale University Press.

Shirky, C. (2011). The political power of social media: Technology, the public sphere, and political change. *Foreign Affairs*, *90*(1), 28–41. https://www.jstor.org/stable/25800379

Shriver, T. E., Adams, A. E., & Longo, S. B. (2015). Environmental threats and political opportunities: Citizen activism in the North Bohemian Coal Basin. *Social Forces*, *94*(2), 699–722. https://doi.org/10.1093/sf/sov072

Snow, D. A., & Benford, R. D. (1988). Ideology, frame resonance and participant mobilization. In B. Klandermans, H. Kriesi, & S. Tarrow (Eds.), *International social movement research* (pp. 197–217). JAI.

Snow, D. A., & Benford, R. D. (1992). Master frames and cycles of protest. In A. D. Morris & C. McClurg (Eds.), *Frontiers in social movement theory* (pp. 133–155). Yale University Press.

Soule, S. A., & Earl, J. (2005). A movement society evaluated: Collective protest in the United States, 1960-1986. *Mobilization: An International Quarterly, 10*(3), 345–364. https://doi.org/10.17813/maiq.10.3.730350353753l022

Strauss, A. L. (1987). *Qualitative analysis for social scientists.* Cambridge University Press.

Tarrow, S. (1998). *Power in movement: Social movements and contentious politics.* Cambridge University Press.

Tilly, C., & Tarrow, S. (2007). *Contentious politics.* Paradigm Press.

Treré, E. (2019). *Hybrid media activism: Ecologies, imaginaries, algorithms.* Routledge.

Tufekci, Z. (2017). *Twitter and tear gas: The power and fragility of networked protest.* Yale University Press.

Uitermark, J. (2017). Complex contention: Analyzing power dynamics within anonymous. *Social Movement Studies, 16*(4), 403–417. https://doi.org/10.1080/14742837.2016.1184136

Van Dijck, J. (2014). Datafication, dataism and dataveillance: Big data between scientific paradigm and ideology. *Surveillance & Society, 12*(2), 197–208. https://doi.org/10.24908/ss.v12i2.4776

Van Dyke, N., & Soule, S. A. (2002). Structural social change and the mobilizing effect of threat: Explaining levels of patriot and militia organizing in the United States. *Social Problems, 49*(4), 497–520. https://doi.org/10.1525/sp.2002.49.4.497

Wills, J. (2017). *Tug of war: Surveillance capitalism, military contracting, and the rise of the security state.* McGill-Queen's University Press.

Wright, S. A. (2007). *Patriots, politics, and the Oklahoma city bombing.* Cambridge University Press.

Zuboff, S. (2019). *The age of surveillance capitalism: The fight for a human future at the new frontier of power.* Profile Books.

# Amplification, evasion, hijacking: algorithms as repertoire for social movements and the struggle for visibility

Emiliano Treré and Tiziano Bonini

**ABSTRACT**

While scholars of activism have begun to unfold the dynamics of the 'contentious politics of data', less explored are the forms of appropriation of algorithms to pursue political objectives by social movements. This article fills this gap by offering a novel theoretical framework, a conceptual vocabulary, and a typology to foreground and articulate algorithmic activism as a subset of algorithmic politics. It starts discussing why an excessive focus on the power of platforms risks disregarding the exploration of agency and provides the definitions of algorithmic agency and politics. Subsequently, it centres on algorithmic activism and demonstrates that algorithms have become the latest addition to the contention repertoire of social movements. Drawing on a heterogeneous set of examples and case studies (including our own research and a database of 250 articles), we propose and examine a typology of three dynamics of algorithmic activism, i.e. algorithmic *amplification*, *evasion*, and *hijacking*. We show that the struggle for visibility (either to achieve it or deny it) lies at the centre of all these types of activism. In the conclusions, we reflect on the key takeaways of our work, clarifying that algorithmic activism (1) exceeds the notion of 'hashtag activism' (2) constitutes an agnostic concept (3) is part of an incessant political struggle between algorithmic strategies and tactics.

## Introduction and outline

Various strands of research have reaffirmed the importance of looking at data systems as battlegrounds of contrasting forces. These studies show that citizens, small-scale organizations, and social movements can make sense of data 'from below' using them for their own needs within their struggles. While scholars of social movements and activism have unfolded the dynamics of the 'contentious politics of data' (Beraldo & Milan, 2019), much less explored are the forms of activist appropriation of algorithms to pursue political objectives. Our contribution fills this gap by critically interrogating the practices that activists and social movements perform to exercise their agency upon algorithms. In our endeavor, we draw on research on data politics (Beraldo & Milan, 2019; Ruppert et al., 2017), social movement studies (Tilly & Tarrow, 2015; Tilly, 2008; Tufekci, 2017; Yang, 2016), resistance studies (Velkova & Kaun, 2021) and algorithm studies (Maly,

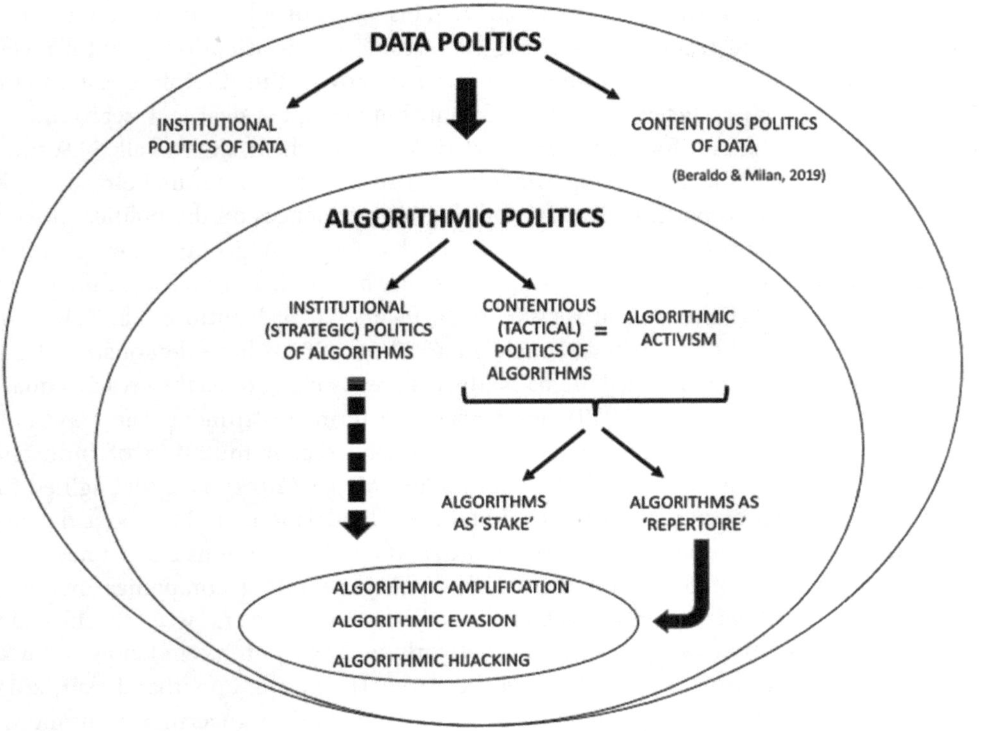


**Figure 1.** From data politics to algorithmic politics: a conceptual framework.

2018, 2019). We build on the emerging scholarship around the political use of data to foreground and articulate the realm of algorithmic politics as a subset of data politics. To accomplish that, we propose and elucidate a novel framework (Figure 1), a related conceptual vocabulary (algorithmic agency > algorithmic politics > algorithmic activism) and a typology of three types of algorithmic activism (algorithmic *amplification, evasion*, and *hijacking*).

The article starts by discussing why an excessive focus on the growing power of platforms risks disregarding the exploration of agency. Then, it offers the definitions of algorithmic agency and algorithmic politics. Subsequently, we focus on algorithmic activism as a form of contentious algorithmic politics and demonstrate that algorithms are the latest addition to the contention repertoire of social movements. Next, drawing on a heterogeneous set of examples and case studies, we propose and examine three types of algorithmic activism, i.e. algorithmic amplification, evasion, and hijacking. We demonstrate that the struggle for visibility (either to achieve it or deny it) lies at the centre of all these types of activism. In the conclusions, we reflect on the key takeaways of our work and its ability to foster dialogue across disciplines.

## The infrastructural turn and the limitations of platform power

The central position acquired by commercial technological giants and their impact on the lives of citizens around the world has propelled an 'infrastructural turn' in media and data studies (Plantin & Punathambekar, 2019). This has spurred a renewed attention to the material aspect of technological infrastructures. Yet, this attention is not a recent 'discovery' of media studies, but rather a 'continuous low' of this discipline (Casemajor, 2015). Scholars such as Peters (2015) have contributed a great deal to foreground the significance of media as infrastructures but, decades before them, the so-called 'Toronto School' (with thinkers as Innis, Ong, McLuhan, Postman and Strate) had already highlighted the material properties of media and their consequences on the political organisation of states. Another benefit of this turn has been to ground the presumed immateriality of 'cyberspace', scrutinizing the social, material, and political dimensions of digital platforms (Parks & Starosielski, 2015; Peters, 2015; Plantin et al., 2018; Van Dijck et al., 2018). Scholars such as Langlois and Elmer (2013) have demonstrated that the realm of data capture of social media platforms greatly expands as they reach a quasi-infrastructural scale. Zuboff (2019) has coined the term 'instrumentarian power' to capture the specific form of power exercised by the recent mutation of industrial capitalism into surveillance capitalism. Couldry and Mejias (2019) have highlighted the colonial continuities of extraction and exploitation of land, labour, and relations through data infrastructure, postulating a new social order where data relations enact a new form of data colonialism. In other words, 'it is becoming clear that companies including Google, Tencent, Amazon, and Facebook that began as platforms with specific aims and areas of operation (shopping, social networking, web search, etc.) now seem to function as vital infrastructures in the world at large' (Plantin & Punathambekar, 2019, p. 164), with disruptive consequences for democracy, citizenship, governance, autonomy, justice, and the environment.

Yet, in realigning our gaze towards how platforms and data systems profoundly reshape media industries, social life and politics, there is a risk of losing sight of the space still available to people to resist the power of these infrastructures. Zuboff (2019) claims that data collection and the use of predictive algorithms by tech industry corporations represent a means of behavioural modification capable of making human behaviour not only completely predictable and manageable, but also automated through a 'digital order that thrives within things and bodies, transforming volition into reinforcement and action into conditioned response' (Zuboff, 2019, p. 378). These accounts may risk reasserting 'monolithic accounts of power that tend to downplay or exclude audiences and the significance of the lifeworld' (Livingstone, 2019, p. 171). We believe that the focus on the increasingly central role of algorithms feeding commercial digital platforms should be combined with the appraisal of the agency that still holds in the hands of citizens in the age of datafication.

## Defining algorithmic agency

The importance of agency in relation to big data has been at the centre of the reflections of scholars who have stressed that datafication should not exclusively be understood as the process of collecting and processing data about users online. This procedure also entails feeding such data back to them, enabling users 'to orient themselves in the world' (Kennedy et al., 2015, p. 1). While big business, governments and corporations often deploy data for surveillance purposes, community groups, small organizations and activist collectives are also leveraging the possibilities of datafication, pursuing objectives, fulfilling needs, and infusing values that are different from those of commercial tech giants (Van Dijck, 2014). Hence, big data constitutes a contested terrain of contrasting forces, where civil society can also produce, gather, and analyse data in ways that enhance the agency of the public, shaping a different kind of datafication. Instead of focusing on top-down processes of datafication, studies on data agency (e.g. Couldry & Powell, 2014), data activism (e.g. Milan, 2017) and everyday practices of 'living with data' (e.g. Kennedy, 2018) dwell on the ways in which different actors make sense of data 'from below' and appropriate them for their own purposes.

It is key to underline that the value of big data originates not so much from the data itself, but from the ways in which it is assembled into new forms of 'meaningfulness by the associational infrastructure of the respective software systems in which algorithms play a key role' (Bucher et al., 2020, p. 94). Algorithms can be seen as 'encoded procedures for transforming input data into desired output, based on specified calculations' (Gillespie, 2014, p. 167). Yet, algorithms are much more than mere computational agents. They are designed to be executed and to bring about particular outcomes according to certain desires, needs, and possibilities' (Willson, 2017, p. 4). They represent 'ethicopolitical arrangements of values, assumptions, and propositions about the world' (Amoore, 2020, p. 6). Data agency and algorithmic agency are intimately connected. Just as data power depends on the ways data are assembled and interpreted, the power of algorithms originates from the choices of who design and apply them. Both data and algorithms respond to certain needs and desires, but they can be appropriated from below to achieve different aims from the ones that were assigned to them. As Bucher argues, 'while algorithms certainly do things to people, people also do things to algorithms' (2017, p. 42). Echoing studies on data agency and activism, scholars have also started to look at algorithmic systems as a battleground of contesting actors, centring on questions of agency and resistance. For instance, Velkova & Kaun questioned 'the extent to which everyday media users are only subjects and victims of algorithmic power' (2021, p. 527) asking if users are powerless against the workings of algorithms. In order to further clarify what we mean by algorithmic agency, it is useful to turn to some definitions. Couldry's understanding of agency underlines the centrality of reflexivity: 'the longer processes of action based on reflection, making sense of the world so as to act within it' (2014, p. 891). Jansen (2016), while distinguishing human agency from the agency of algorithms, considered the former an evaluative and reflexive action and claimed that algorithms lack evaluative and reflexive agency. We are aware of the ongoing conversation regarding the differences between human and non-human agency in relation to algorithms (Klinger & Svensson, 2018). However, combining the above definitions, with the term *algorithmic agency* we hereby refer not to the agency of the algorithms but rather to the user's 'reflexive ability' to make the algorithms work to their own needs.

## Defining algorithmic politics

As we saw in the previous section, data and algorithmic agency can be exercised from above and from below for different purposes. But what happens when we move to the realm of politics? Ruppert et al. (2017, p. 1) argue that 'data politics is concerned with the conditions of possibility of data that involve things (. . .), language (. . .), and people (. . .)

that together create new worlds'. They then define 'data politics' as 'both the articulation of political questions about these worlds and the ways in which they provoke subjects to govern themselves and others by making rights claims' (2017, p. 1), specifying that it is 'concerned with not only political struggles around data collection and its deployments, but how data is generative of new forms of power relations and politics at different and inter-connected scales' (2017, p. 2). Expanding on this definition of data politics and building on the notion of 'contentious politics' by social movement scholars Tilly and Tarrow (2015), Beraldo and Milan (2019) have introduced a differentiation between the 'institutional politics of data' - which refers to the top-down effects of datafication on groups and individuals – and the 'contentious politics of data' - which refers to 'the bottom-up practices embodied and promoted by individuals and groups' (p. 4). With contentious politics of data, the two scholars refer to 'the multiplicity of bottom-up, transformative initiatives interfering with and/or hijacking dominant, top-down processes of datafication, by contesting existing power relations and narratives and/or by re-appropriating data practices and infrastructure for purposes distinct from the intended' (2019, p. 2). In the wake of this distinction made by Beraldo & Milan, we propose a similar distinction for what we call *algorithmic politics*, a subset of data politics concerned with how different groups exercise their algorithmic agency by appropriating and acting upon algorithms to fulfil their political objectives (see Figure 1 in the conclusions). We differentiate between an '*institutional* algorithmic politics' and a '*contentious* algorithmic politics. The former refers to the attempts to act on algorithms undertaken from 'above' - by a state, an institution, a corporation, etc. And is often referred to as 'computational propaganda' (Woolley & Tucker, 2020) or 'algorithmic populism' (Maly, 2018, 2019). The latter denotes instead all those practices initiated from 'below' – by a collective, a social movement or individual activists and has been addressed as 'algorithmic resistance' (Treré, 2018, 2019) or 'algorithmic activism' (Maly, 2018, 2019). The latter type is the focus of this article and is detailed in the next section.

## *Contentious and tactical algorithmic politics (algorithmic activism)*

Datafication and algorithmically mediated environments are restructuring collective action and the dynamics of social movements at a profound level (Milan, 2015), changing the very conditions under which social movements operate (Beraldo & Milan, 2019; Flesher Fominaya, 2020; Treré, 2016, 2019). The algorithms of social media platforms have several implications for the activities of activists. Algorithms represent key actors in what Dolata (2017) has called the 'socio-technical constitution of collective action'. Maly (2018, 2019) has coined the concept of 'algorithmic activism' to tackle the rise of a Flemish far-right activist movement. In his analysis, Maly sheds light on how 'algorithmic activists' from this movement strategically exploit the affordances of social media to reach their goals, 'boost their popularity rankings' and make their content go viral (Maly, 2019, p. 1). In this article, we identify *algorithmic activism* with the contentious politics of data that is concerned with how a range of actors including activists, social movements and civil society actors engage with and act upon algorithms to achieve their political aims and pursue social change. We therefore conceptualise this phenomenon as a form of contentious algorithmic politics exercised through a tactical form of algorithmic agency.

## Algorithms as repertoire and the struggle for visibility

Tilly originally defined a 'repertoire of contention' as including the 'whole set of means [a group] has for making claims of different types on different individuals' (1986, p. 2). He later clarified that this notion refers to 'claims making routines that apply to the same claimant-object pairs: bosses and workers, peasants and landlords, rival nationalist factions, and many more' (Tilly, 2010, p. 35). While Tilly's original conceptualisation has been criticized for its narrow focus on public display of disruptive action, more recently (2008) he has extended its focus to include broader contentious performances, underlining the constant innovation of contentious politics. Reflecting on the increasing relevance that new media have acquired within the repertoires of social movements, scholars have started to use labels such as 'electronic repertoire of contention' (Rolfe, 2005) and 'digital network repertoires' (Breindl & Briatte, 2010). Both social movement and media scholars have adopted repertoires of contention to address how activists integrate digital technologies as a set of means to challenge authority, protest, and mobilise (Liu, 2020). In the realm of data activism, Beraldo and Milan (2019) have introduced a distinction between 'data as stakes' and 'data as repertoire'. In the latter type (data-enabled activism), data are inserted within the repertoire of action of social movements and activists 'alongside other more traditional forms of protest and civic engagement' (Beraldo & Milan, 2019, p. 6). Inspired by this distinction, in this article we focus on citizen practices that deploy algorithms as repertoire (algorithm-enabled activism), including the creativity, the resourcefulness, and the difficulties that activists face while coping with opaque decisions taken by an algorithm. Algorithms constitute the latest addition to the repertoire of contention of contemporary protest.

Since most of the digital environments that activists navigate are algorithmically determined, the capacity to understand, adapt and re-purpose algorithms lies at the centre of today's collective action. Different scholars have touched upon this topic from disciplines as varied as computer science, political science, media, data, algorithm, management, resistance, and social movement studies. For instance, scholars working at the intersection between social movement and media studies have addressed the ways in which activists and algorithms are mutually entangled in protest-related situations. Galis and Neumayer (2016), in their study of digital protest in Greece and Sweden, have introduced the concept of 'cyber-material détournement'. With this notion, they refer to the alliances between activists and non-human actors (with algorithmic agents *in primis*) that define social media activism. Other thinkers working at the intersection between resistance and data studies have grown dissatisfied with dystopian accounts of algorithmic power and have started to rely on De Certeau's understanding of tactics (1984). Velkova & Kaun, for instance, 'foreground the significance of mundane user encounters with algorithms through which users can develop tactics of resistance through alternative uses' (2021, p. 525). Relatedly, authors such as Gangneux (2020) talk of 'tactical agency' to describe the ways in which young people engage (and disengage) with WhatsApp and Facebook Messenger. Kant (2020) describes users who are engaged in 'manoeuvring within, against and through algorithmic anticipation' (2020, p. 215) as 'algorithmic tacticians'. Social movements engage in algorithmic activism by leveraging the opportunities offered by social media's algorithms to various ends. Primarily, they do it to attract the attention of mainstream media by increasing their visibility. Visibility can be understood in this context as the 'digital embodiment and online presence of individuals and groups and their associated meanings, which are (. . .) constantly negotiated, reinvigorated, and updated' (Milan, 2015, p. 6). Algorithmic visibility is a type of power (Bucher 2018) that selects who and which content can or cannot be seen. In algorithmically determined settings, activists are pushed even more to play the game of visibility (Cotter, 2019). But alongside visibility, they also appropriate algorithms to exercise and fuel their 'narrative agency' (Yang, 2016), that is the capacity to tell their stories and frame events and experiences in their own terms. This is testified by movements as diverse as #OccupyWallStreet, #BlackLivesMatter and #YoSoy132. As social movement scholar Zeynep Tufekci has underlined (2017), the strongest movements are the ones that are able to develop a 'narrative capacity', understood as the ability to attract public attention and introduce new issues or frames into the political debate. In the datafied society, algorithmic activism represents a key component in the construction of this capacity for many contemporary movements.

## The design of the typology of algorithmic activism

In the following section, we build on our own work, on the work of other scholars, and on examples from global socio-political contexts to illustrate the variety of practices in which algorithms are used as repertoire for contemporary movements and activists. We outline how different types of algorithmic activism centre on a struggle around voice and the incessant quest for visibility. This entails attempts to achieve that a social movement's voice is recognized, and its visibility strengthened, while at the same time it includes the negation of other social actors' visibility and the attempts to silence their voice. We derive our ideas from a heterogenous set of examples, including cases from our own research. Based on these examples, we propose a typology of three types of algorithmic activism: algorithmic *amplification, evasion, and hijacking.*

Our own research carried out in Mexico and in Spain within two different research projects on social movements and digital media represents the first source for the typology construction. During 2017 and 2018, we carried out fifteen expert interviews with a selection of prominent Mexican data activists, human rights advocates, journalists, politicians, media strategists and academics. In the case of Spain, the ideas regarding the forms of algorithmic amplification of the 15 M movement draw on 20 semi-structured interviews of an hour and a half each with social movement and media activists carried out in 2015 in Madrid, Barcelona, and Seville. A snow-ball sampling strategy was used to select twenty participants covering a range of social movement organisations. The interviews addressed people who played key roles in organising and producing media about the anti-austerity protests, including journalists, web managers and developers, social media curators, graphic designers, media activists, and precarious media researchers. Both projects involved the technique of 'active interviewing' developed by Holstein and Gubrium (1995) that uses broad questions to give agency to research participants. Such a technique, based on a conception of reality as an ongoing and interpretative accomplishment, is consistent with a conception of the relationships between activists and algorithms as a changing, dynamic complex. Respondents could address a wide range of meanings by telling stories in response to wide-ranging questions, such as 'How would you describe your relationship with this platform?' or 'In which ways would you define your engagement with the algorithm of Twitter?'. This allowed us to foreground the 'narrative capacity' of activists in their engagement with communication technologies (Atkinson & Cooley, 2010, p. xiv). The second source of examples for the typology construction emerges from desk research on both scientific papers and the press around algorithm 'gaming' (cheating) practices.

For the selection of this sample, we first extracted relevant keywords from a set of 10 articles concerning cases of 'algorithmic gaming' or 'cheating' in the context of gig working, political activism, and cultural production, which we had read independently. The most relevant words, such as 'algorithmic gaming', 'algorithmic scams', 'cheating algorithms', 'optimization', 'amplification', 'resistance', 'solidarity', and 'algorithmic manipulation' have been paired with 'activism', 'gig working' and 'cultural production' and used as keywords in Google, Twitter and Google Scholar search engines. From this search we build a database made up of 100 press articles and 150 scientific articles, whose publication dates range from 2013 to 2021. We carried out a content analysis on this database and coded the press articles according to three categories: 'gaming culture', 'gaming politics', and 'gaming work', while the scientific articles were coded according to their relevance to one of the previous categories and the themes of agency and resistance. For this article, we selected key examples from the 'gaming politics' category.

## Algorithmic amplification

In specific political junctures, activists can integrate algorithms into their repertoire to multiply and amplify their voice to acquire more visibility while strengthening their agency and narrative capacity. For example, scholars from political science have addressed the use of algorithms on Twitter to pursue social change and insert alternative narratives through the notion of 'hashtag activism' (Jackson et al., 2020). These scholars examine how marginalized groups rely on hashtags on Twitter (e.g. #MeToo, #GirlsLikeUs, and #BlackLivesMatter) to pre-empt political spin, build networks of dissent and challenge dominant understandings of gender and race. In our own work (Treré, 2018, 2019), we have mobilized the notion of 'algorithmic resistance' to characterize various strategies of appropriation of social media algorithms by protest movements to pursue their political aims and achieve greater visibility. This concept relates specifically to the technopolitical practices of the Spanish movement 15 M. Through the tactical adoption of social media and their algorithms, the movement was able to spread information, organise protests, build powerful narratives and shape both national and international journalistic coverage. One of the most representative tactics of algorithmic activism established by Spanish activists consisted of the prearranged creation of trending topics on Twitter (see also Feenstra et al., 2017). This was a carefully planned tactic that comprised a blend of internal communication technologies and corporate social media. Tools of internal communication such as online pads (digital notepads for collective writing), were used – in the words of an activist – 'to collectively select a list of possibly successful hashtags and build the narrative of the protest', while 'external' social media platforms (Twitter in particular) - were instead deployed to massively spread the information and reach the desired result. Inside the pads, the most effective hashtags were brainstormed and selected depending on the activist campaign that was undertaken. Once a hashtag was chosen, a range of potential tweets was then created accordingly and shared with other activists through internal communication environments – a variable combination of direct messages on Twitter, instant messaging services (WhatsApp, Telegram, Signal) and mailing lists. This refined tactic presupposes a deep awareness and reflective knowledge of how algorithms operate. As one 15 M activist explains, this was 'obtained through incessant sequences of trial and error', originating from 'trying to understand how the Twitter algorithm worked and how we could exploit it for improving our visibility and influence the mainstream media agenda'. As another activist adds:

> Our aim was to hack the Twitter algorithm so our narratives, our voices, our ways of seeing things could reach as many people as possible. We use corporate social media because this is where most people are, and we want to reach the highest number of persons out there. We want to be visible, and we want our message to get across and be picked up by other mainstream media as well. We know that social media are extractive and use our data in many ways, but this time it's us using them to multiply our presence and empower the movement.

Spanish activists made a series of 'discoveries' in relation to the Twitter algorithm. They found that the general trending topics had a cycle of 24 hours and that 'all the accounts needed to tweet simultaneously with the same hashtag' (Interview with a media activist). Furthermore, they noticed the hashtag had to be 'fresh' – i.e. not used before – since the Twitter algorithm tends to always reward newness.

In other cases, activists can also rely on algorithms to amplify the visibility of someone to expose them with the objective of hurting their public image. For example, in 2018, mainly in response to Donald Trump's policies on immigrations and LGBTQ rights, activists started to manipulate Google' search algorithm by massively linking the word 'idiot' to pictures of him (Haynes, 2018). The association between the word 'idiot' and the former president of the USA was partly ignited by the choice of London protesters of the Green Day song 'American Idiot' during the Trump's visit to England. Activists on Reddit started to upvote posts of Trump[1] in connection to the word 'idiot' lending Google's ranking algorithm to associate the two. Google's position in relation to this incident has been not to interfere with its search results. The company has a long history of this type of scandals, with offensive and sometimes racist content being linked to particular individuals. Another illustration of this happened in 2009 when searches for 'Michelle Obama' returned a picture of the first lady's face with ape-like features. These controversies highlight the inherent biases of algorithmically driven decisions (Noble, 2018). Further, they exemplify the evolving and contested dynamics of algorithmic power at the intersection between the interests of digital platforms, political figures, activists, and the public. We have seen how activists can appropriate algorithms to achieve more visibility for themselves or to amplify someone else's biased voice to hurt their public image. Yet, activists can also repurpose algorithmic power to repair what they feel is an unjust form of invisibility, thus preventing someone else's voice from being silenced or unheard. In relation to this, Velkova and Kaun (2021) have focused on forms of explicit algorithmic resistance though their concept of 'media repair practices'. These 'media practices of repair are tactics to correct existing shortcomings *within* algorithmic culture rather than by producing alternative pathways. In that sense, they establish reactive user agency in an algorithmic aftermath' (Velkova & Kaun, 2021, pp. 523–4). The two authors explore the 'World White Web' project by Swedish design student Johanna Burai. The artist tried to tweak Google's image search algorithm to disseminate alternative search results, after discovering that her basic photographic search of a human hand returned almost exclusively images of white hands. In 2015, Burai launched a campaign to get six images of non-white hands as top results in Google image searches. Hence, the activist appropriation of algorithms as repertoire is key also to respond to emerging forms of algorithmic injustices.

## Algorithmic evasion

Examples of activists fighting against digital platforms' codes and algorithmic regulations to have their voices heard include the realm of 'content moderation avoidance strategies'. These strategies have been observed in different domains as varied as vaccine-opposed groups (Moran et al., 2021), Pro-Eating Disorder communities (Chancellor et al., 2016; Gerrard, 2018), and far-right movements (Bhat & Klein, 2020). In relation to the first type, Moran et al. (2021) have illustrated that COVID-19 vaccine-opposed groups can bypass community guidelines and moderation features on social media by using various expedients. One of the most diffused is lexical variation on Twitter, where iterations of words – e.g. v@ccine – are used to prevent algorithmic detection and blocking of content. On the same platform, they would also create vaccine-neutral hashtags to spread misinformation within pro-vax conversations. As Moran et al. (2021) demonstrate, the practices of anti-vax strategists are always carefully adapted to the specific digital platform they are relying on. On Facebook, they would provide links to vaccination misinformation in the comment section, rather than in the original post. On Instagram, they would develop ephemeral content strategies including anti-vaccination content on stories rather than in-feed or rely on coded language using words like 'toxins' and 'metals' instead of vaccines.

In the case of Pro-Eating Disorder (Pro-ED) groups on Instagram, similar strategies of content moderation circumvention were detected in response to Instagram's (and other digital platforms) ban of phrases such as 'anorexia', 'proana (pro-anorexia)', 'thinspiration', 'thighgap' and 'imugly' and the blocking of the results of certain hashtag searches since 2012. This has led to the proliferation of lexical variants of the banned tags to deceive algorithmic control, with for example 'thighgap' replaced with 'thyghgapp' and 'thinspo' turning into 'thinspoooooo'. Research (Chancellor et al., 2016) has shown how the Instagram's ban worsened the situation with new variants having an even further reach than those they were designed to replace, spreading to other platforms such as Tumblr and Twitter.

Similarly, Bhat and Klein (2020) have illustrated how far-right activists engaged in practices of 'dog whistling' to evade censorship by automated moderation on Twitter during the 2016 US presidential election. They draw on the notion of 'dog whistling' to denote the use of symbols and terminology that means something to the larger audience, but acquires a different meaning for a specific public, in this case the white supremacist audience. Words as 'googles' or 'Dindu Nuffins' for African Americans and 'skittles' to indicate Muslims, together with other symbols such as parenthesis and percentages were used to spread hate and racist, xenophobic content that escaped Twitter's algorithmic detection mechanisms. Relatedly, the reliance on a particular language for avoiding algorithmic control and surveillance under authoritarian governments has been observed in contemporary Turkey. As Tonnesen (2020) points out, activists who oppose the government appropriate the 'vernacular Twitter language and the Western popular culture which includes drawing subtle comparisons between movie villains and the President, references to early Internet phenomena, captioning viral videos with implied political messages etc' (p. 2–3). This allows them to have their voice heard while flying under the radar of Turkish government's social media control, since this playful vernacular language is out of sync with pro-government forces' digital skills. These cases prove that algorithmic activism is not only about appropriating algorithms to amplify visibility and strengthen narrative agency, but also about finding new ways to evade them to avoid being silenced by digital platforms.

## Algorithmic hijacking

With the notion of 'hashtag hijacking', the computer science literature designates a practice where hashtags are used to spread un-related content, spam, or negative sentiments. The objective is to tarnish the intended motive of a hashtag, thus rendering its presence counterproductive (Jain et al., 2015, p. 17). A powerful example of this tactic is represented by the collective hijacking of the Twitter hashtag #myNYPD described by Jackson and Foucault Welles (2015) that followed the launch of a public relations campaign by the New York City Police Department in 2014. The authors document how activists' hijacking of the police hashtag led to the formation of a counter-public sphere. They use this experience to explain how the repertoire of digital contention of citizen activists is evolving in the algorithmic age. More recently, K-pop fandoms – known for their dedication to the 'idols' of South Korean music – have used their sophisticated skills in the art of algorithmic manipulation and hijacking at the service of social justice causes. In 2020, protests emerged in response to the police killing of George Floyd, and the subsequent threat of former US president Trump to deploy the army against protesters. In this contest, the Dallas Police Department asked the public to submit video of 'illegal activity from the protests' through a dedicate app. K-pop fans inundated the app with K-pop-related content, in particular fancams (fan-edited videos of K-pop stars). As a result, the Dallas Police Department were forced to take down the app (Romano, 2020). K-pop fans have also hijacked right-wing hashtags like #MAGA, #BlueLivesMatter and #whitelivesmatter flooding them with clips of K-pop groups, memes and other similar content to drown out racist and offensive voices. As Abidin and Baudinette (2020) have pointed out, the fact that K-pop fans are mastering these algorithmic tactics is no surprise to fandom scholars who have been following their practices for years. These communities have a long history of 'being political', relying on online platforms for social justice purposes, charitable causes and online vigilantism (Ohlheiser, 2020). Their high level of tech-savviness and algorithmic awareness is born out of their constant interactions with the affordances of social media. Finally, algorithmic hijacking is not merely related to hashtags. For example, police officers in the US have been playing copyrighted music with their phones during attempts to be recorded by activists in the course of their encounters. Their aim is to prevent those videos to be posted on YouTube, and to silence activists' voices avoiding accountability. To do so, they exploit the platform's copyright policy that filters and remove this kind of material through a system called

Content ID (Sung, 2021). This kind of algorithmic hijacking shows the hybrid nature of the phenomenon that can take place at the intersection between the physical and the digital realms.

## The struggle between algorithmic strategies and tactics

So far, we have mainly focused on manifestations of algorithmic activism. In this last section, we instead zoom in on a recent example from the contemporary Mexican political scenario that perfectly encapsulates a form of institutional/strategic algorithmic hijacking. Our aim is to show that (1) the three mechanisms (amplification, evasion, hijacking) that we have outlined can be used from above by institutional actors and are not the exclusive realm of citizens and activists; (2) algorithmic politics is characterized by an incessant back-and-forth struggle between algorithmic strategies and tactics.

On 26 September 2014, six deaths and the forced disappearance of 43 students at the Ayotzinapa teachers' college in Guerrero (Mexico) spurred the emergence of a social movement in solidarity with the families of the victims whose main aim was 'to present the missing students' lives' (Gravante, 2020, p. 88). The spark of the movement was ignited after the event, when several outraged activists started to protest on social media, and the Twitter hashtag #YaMeCanse (IAmTired) – which expressed the feeling of not being able to tolerate any more violence – soon became a key space for protest organizing, information spreading and one of the most used hashtags in Mexican history. Multimedia artist and writer Erin Gallagher has collected a detailed database of bot attacks and algorithmic strategies in Mexico and beyond (Gallagher, 2017). In relation to #YaMeCanse, she noticed something atypical in her search results for the hashtags: they were flooded with tweets, including the hashtag, but no other content besides random punctuation marks. The accounts tweeting this kind of empty content were, in fact, bots with no followers and they were tweeting automatically. As documented by the Mexican algorithmic activist Escorcia (2014), automated accounts purposively hijacked the hashtag with 'noise', inserting links to pornography, advertisements, and violent images, so the Twitter algorithm would flag it as spam and consequently block it. The confusion they generated made it difficult for citizens to share and rely on information through #YaMeCanse. Hence, the hashtag vanished from Twitter's list of trending topics. However, activists reacted to this disruption by producing different lexical iterations of it such as 'YaMeCanse1' and 'YaMeCanse2'. These iterations were able to move the conversation elsewhere eluding the noise created by bots probably fuelled by the local government. These last tactics are a further example of how algorithms can be used to evade their own control, as we have seen in the previous section.

This example shows how institutions can deploy strategies of algorithmic hijacking to prevent activists' voice from being heard and silence dissent online. However, activists can leverage social media algorithms to react to these institutional strategies promptly. Through algorithmic evasion and amplification, they can reclaim spaces for their own narratives in the never-ending battle between algorithmic strategies and tactics that defines the algorithmic age.

## Discussion and conclusion

Relying on contributions from data and social movement studies, this article identified the field of algorithmic politics as a sub-set of data politics, concerned with how different groups appropriate algorithms to fulfil their political objectives. We then introduced the distinction between institutional and contentious uses of algorithms, focusing on the latter type which exemplifies the very field of algorithmic activism. Within this type of activism, we saw that algorithms can be the tools *through* which activists exercise their protest (algorithms as repertoire), representing the latest addition to the repertoire of contention of protest movements. We then introduced and discussed three types of algorithmic activism: algorithmic amplification, evasion, and hijacking. Figure 1 shows the conceptual framework which represents the main contribution of this article.

This framework brings together the scattered pieces of research that have dealt with emerging forms of political engagement with algorithms from various perspectives and by a diverse range of actors. It foregrounds, connects, and articulates all the facets of the contested battleground of algorithmics politics, conceived as a subset of data politics.

Moreover, there are three additional contributions of our paper. Firstly, it has become evident that our conception of algorithmic activism includes but exceeds the notion of 'hashtag activism'. Algorithmic activism operates not only through hashtags, but over a broader variety of platforms, devices, actors, and socio-political domains. While Twitter has been a key environment for the unfolding of recent protests, algorithmic activism encompasses a wider spectrum of tactics, dynamics, and subterfuges. As algorithms increasingly pervade more aspects of our lives, these practices are becoming more mundane and integrated into the fabric of our everyday experiences. For example, management scholars Kellogg, Valentine & Christin use the similar term 'algoactivism' to address the individual and collective tactics of gig workers in their resistance efforts against algorithmic control (2020).

Secondly, another important takeaway relates to the *agnosticism* of algorithmic activism. As the cases we have mapped testify, algorithms are being equally appropriated by racist and xenophobic movements, oppressive regimes as well as by progressive activists and radical collectives. The political terrain of this activism is multifaceted and contradictory (Etter & Albu, 2021), with various forces at play engaged in a struggle to either obtain or thwart visibility. Thirdly, through the example of institutional/strategic hijacking in Mexico and the reactions it engendered in activists' tactics, we have illustrated how the field of algorithmic politics is crossed by an incessant struggle between algorithmic strategies and tactics. The three algorithmic dynamics outlined in this article can thus be appropriated both from above by institutional actors and from below by activist tacticians. We hope that this framework will be able to foster a dialogue (and generate further research) across the disciplinary boundaries of critical data and algorithm studies and social movement studies. As the processes of datafication and automation through algorithms become pervasive in our daily lives, data and algorithms do not only represent instruments in the hands of institutions and corporations. They also become part of the repertoire that citizens and social movements embrace to exercise their collective and individual agency. We believe that the ideas discussed in this article can enrich the conceptual vocabulary of social movement studies in their engagement with the complexities of politics in the age of datafication.

## Note

1. See for instance: https://www.reddit.com/r/The_Mueller/comments/8on76n/the_first_picture_that_comes_up_when_you_google/.

## Acknowledgment

We are grateful to the School of Journalism, Media and Culture (JOMEC) at Cardiff University (Wales, UK) for providing funding that supported the creation of the database used in this article through the AlgoRes Project. A special thanks to Thomas Davis for his hard work and dedication.

## Disclosure statement

No potential conflict of interest was reported by the author(s).

## Funding

The work was supported by the Consejo Nacional de Ciencia y Tecnología [SNI (Mexican National System of Researchers)]; School of Journalism, Media and Culture (JOMEC), Cardiff University, Wales, UK [Seed Funding]; Social Sciences and Humanities Research Council of Canada [430-2014-00181]

## ORCID

Emiliano Treré http://orcid.org/0000-0002-2496-4571
Tiziano Bonini http://orcid.org/0000-0002-0636-0555

## References

Abidin, C., & Baudinette, T. (2020 , July 1) The civic Hijinks of K-Pop's super fans, data & society. https://points.datasociety.net/the-civic-hijinks-of-k-pops-super-fans-ae2e66e28c6.

Amoore, L. (2020). *Cloud ethics*. Duke University Press.

Atkinson, J. D., & Cooley, L. (2010). Narrative capacity, resistance performance, and the "shape" of new social movement networks. *Communication Studies*, *61*(3), 321–338. https://doi.org/10.1080/10510971003752668

Beraldo, D., & Milan, S. (2019). From data politics to the contentious politics of data. *Big Data & Society*, *6*(2), 2053951719885967. https://doi.org/10.1177/2053951719885967

Bhat, P., & Klein, O. (2020). Covert hate speech: White nationalists and dog whistle communication on Twitter. In G. Bouvier & J. E. Rosenbaum (Eds.), *Twitter, the public sphere, and the chaos of online deliberation* (pp. 151–172). Palgrave Macmillan.

Breindl, Y., & Briatte, F. (2010). Digital network repertoires and the contentious politics of digital copyright in France and the European Union. *Internet, Politics, Policy 2010: An Impact Assessment*.

Bucher, T. (2017). The algorithmic imaginary: Exploring the ordinary affects of Facebook algorithms. *Information, Communication & Society*, *20*(1), 30–44. https://doi.org/10.1080/1369118X.2016.1154086

Bucher, T. (2018). *If. then: Algorithmic power and politics*. Oxford University Press.

Bucher, T. (2020) (Big) data and algorithms. Looking for meaningful patterns. In Lievrouw, L. A. & Loader, B. D. (eds), *Routledge handbook of digital media and communication* (pp. 87–98). Routledge.

Casemajor, N. (2015). Digital materialisms: Frameworks for digital media studies. *Westminster Papers in Communication and Culture, 10*(1), 4–17. https://doi.org/10.16997/wpcc.209

Chancellor, S., Pater, J., Clear, T., Gilbert, E., & De Choudhury, M. (2016). #thyghgapp: Instagram content moderation and lexical variation in pro-eating disorder communities. *Georgia Institute of Technology*. 1. http://www.munmund.net/pubs/cscw16_thyghgapp.pdf

Cotter, K. (2019). Playing the visibility game: How digital influencers and algorithms negotiate influence on Instagram. *New Media & Society, 21*(4), 895–913. https://doi.org/10.1177/1461444818815684

Couldry, N. (2014). Inaugural: A necessary disenchantment: Myth, agency and injustice in a digital world. *The Sociological Review, 62*(4), 880–897. https://doi.org/10.1111/1467-954X.12158

Couldry, N., & Mejias, U. A. (2019). *The costs of connection: How data is colonizing human life and appropriating it for capitalism*. Stanford University Press.

Couldry, N., & Powell, A. (2014). Big data from the bottom up. *Big Data & Society, 1*(2), 2053951714539277. https://doi.org/10.1177/2053951714539277

De Certeau, M. (1984). *The practice of everyday life*. University of California Press.

Dolata, U. (2017). *Social movements and the internet: The sociotechnical constitution of collective action (No. 2017-02)*. SOI Discussion Paper.

Escorcia, A., (2014). *LoQueSigueTV*. https://loquesigue.tv/

Etter, M., & Albu, O. B. (2021). Activists in the dark: Social media algorithms and collective action in two social movement organizations. *Organization, 28*(1), 68–91. https://doi.org/10.1177/1350508420961532

Feenstra, R. A., Tormey, S., Casero-Ripollés, A., & Keane, J. (2017). *Refiguring democracy: The Spanish political laboratory*. Routledge.

Flesher Fominaya, C. F. (2020). *Democracy reloaded: Inside Spain's Political laboratory from 15-M to podemos*. Oxford University Press.

Galis, V., & Neumayer, C. (2016). Laying claim to social media by activists: A cyber-material détournement. *Social Media+ Society, 2*(3), 2056305116664360. https://doi.org/10.1177/2056305116664360

Gallagher. (2017). Mexico: articles about bots and trolls. https://erin-gallagher.medium.com/news-articles-about-bots-trolls-in-mexican-networks-7b1e551ef4a6

Gangneux, J. (2020). Tactical agency? Young people's (dis) engagement with WhatsApp and Facebook Messenger. *Convergence, 1354856520918987*(2), 458–471. https://doi.org/10.1177/1354856520918987

Gerrard, Y. (2018). Beyond the hashtag: Circumventing content moderation on social media. *New Media & Society, 20*(12), 4492–4511. https://doi.org/10.1177/1461444818776611

Gillespie, T. (2014). The relevance of algorithms. *Media Technologies: Essays on Communication, Materiality, and Society, 167*(2014), 167.

Gravante, T. (2020). Forced disappearance as a collective cultural trauma in the ayotzinapa movement. *Latin American Perspectives, 47*(6), 87–102. https://doi.org/10.1177/0094582X20951773

Haynes, G. (2018, July 17). Search 'idiot' get trump: How activists are manipulating Google images. *The Guardian*. https://www.theguardian.com/us-news/2018/jul/17/trump-idiot-google-images-search

Holstein, J. A., & Gubrium, J. F. (1995). *The active interview* (Vol. 37). Sage.

Jackson, S. J., Bailey, M., & Welles, B. F. (2020). *# HashtagActivism: Networks of race and gender justice*. MIT Press.

Jackson, S. J., & Foucault Welles, B. (2015). Hijacking# myNYPD: Social media dissent and networked counterpublics. *The Journal of Communication, 65*(6), 932–952. https://doi.org/10.1111/jcom.12185

Jain, N., Agarwal, P., & Pruthi, J. (2015). HashJacker-detection and analysis of hashtag hijacking on Twitter. *International Journal of Computer Applications, 114*(19), 17–20. https://doi.org/10.5120/20085-2111

Jansen, T. (2016). Who is talking? Some remarks on nonhuman agency in communication. *Communication Theory, 26*(3), 255–272. https://doi.org/10.1111/comt.12095

Kant, T. (2020). *Making it personal: Algorithmic personalization, identity, and everyday life*. Oxford University Press.

Kellogg, K. C., Valentine, M. A., & Christin, A. (2020). Algorithms at work: The new contested terrain of control. *The Academy of Management Annals*, *14*(1), 366–410. https://doi.org/10.5465/annals.2018.0174

Kennedy, H. (2018). Living with data: Aligning data studies and data activism through a focus on everyday experiences of datafication. *Krisis: Journal for Contemporary Philosophy*, 1, 18-30.

Kennedy, H., Poell, T., & van Dijck, J. (2015). Data and agency. *Big Data & Society*, *2*(2), 1–7. https://doi.org/10.1177/2053951715621569

Klinger, U., & Svensson, J. (2018). The end of media logics? On algorithms and agency. *New Media & Society*, *20*(12), 4653–4670. https://doi.org/10.1177/1461444818779750

Langlois, G., & Elmer, G. (2013). The research politics of social media platforms. *Culture Machine*, *14*, 1-17.

Liu, J. (2020). *Shifting dynamics of contention in the digital age: Mobile communication and politics in China*. Oxford University Press.

Livingstone, S. (2019). Audiences in an age of datafication: Critical questions for media research. *Television & New Media*, *20*(2), 170–183. https://doi.org/10.1177/1527476418811118

Maly, I. (2018, October 28). Algorithmic populism and algorithmic activism. *Diggit Magazine*. https://www.diggitmagazine.com/articles/algorithmic-populism-activism

Maly, I. (2019). New right metapolitics and the algorithmic activism of Schild & Vrienden. *Social Media+ Society*, *5*(2), 2056305119856700. https://doi.org/10.1177/2056305119856700

Milan, S. (2015). When algorithms shape collective action: Social media and the dynamics of cloud protesting. *Social Media+ Society*, *1*(2), 2056305115622481. https://doi.org/10.1177/2056305115622481

Milan, S. (2017). Data activism as the new frontier of media activism. In V. Pickard & G. Yang (Eds.), *Media activism in the digital age* (pp. 151–163). Routledge.

Moran, R., Koltai, K., Grasso, I., Schafer, J., Klentschy, C. (2021) Content moderation avoidance strategies, virality project. https://www.viralityproject.org/rapid-response/content-moderation-avoidance-strategies-used-to-promote-vaccine-hesitant-content

Noble, S. U. (2018). *Algorithms of oppression: How search engines reinforce racism*. NYU Press.

Ohlheiser, A. (2020, June 5). How K-pop fans became celebrated online vigilantes, MIT Technology Review. https://www.technologyreview.com/2020/06/05/1002781/kpop-fans-and-black-lives-matter/

Parks, L., & Starosielski, N. (Eds.). (2015). *Signal traffic: Critical studies of media infrastructures*. University of Illinois Press.

Peters, J. D. (2015). *The marvelous clouds: Toward a philosophy of elemental media*. University of Chicago Press.

Plantin, J. C., Lagoze, C., Edwards, P. N., & Sandvig, C. (2018). Infrastructure studies meet platform studies in the age of Google and Facebook. *New Media & Society*, *20*(1), 293–310. https://doi.org/10.1177/1461444816661553

Plantin, J. C., & Punathambekar, A. (2019). Digital media infrastructures: Pipes, platforms, and politics. *Media, Culture & Society*, *41*(2), 163–174. https://doi.org/10.1177/0163443718818376

Rolfe, B. (2005). Building an electronic repertoire of contention. *Social Movement Studies*, *4*(1), 65–74. https://doi.org/10.1080/14742830500051945

Romano, A. (2020, June 22) How K-pop fans are weaponizing the internet for black lives matter, Vox.Com. https://www.vox.com/2020/6/8/21279262/k-pop-fans-black-lives-matter-fancams-youtubers-protest-support

Ruppert, E., Isin, E., & Bigo, D. (2017). Data politics. *Big Data & Society*, *4*(2), 2053951717717749. https://doi.org/10.1177/2053951717717749

Sung, M. (2021 , July 1) Cops are playing music during filmed encounters to game YouTube's copyright striking. But will it work? Mashable. https://mashable.com/article/police-playing-music-copyright-youtube-recording

Tilly, C. (1986). *The contentious French: Four centuries of popular struggle*. Belknap Press.

Tilly, C. (2008). *Contentious performances*. Cambridge University Press.

Tilly, C. (2010). *Regimes and repertoires.* University of Chicago Press.

Tilly, C., & Tarrow, S. (2015). *Contentious politics* (2nd ed.). Oxford University Press.

Tonnesen, N. O. D. (2020). "Two can play at that game": Communicating dissent as a micro-celebrity in a restricted national twittersphere". *AoIR Selected Papers of Internet Research.* https://doi.org/10.5210/spir.v2020i0.11200

Treré, E. (2016). The dark side of digital politics: Understanding the algorithmic manufacturing of consent and the hindering of online dissidence. *IDS Bulletin, 47*(1), 127–138. https://doi.org/10.19088/1968-2016.111

Treré, E. (2018). From digital activism to algorithmic resistance. In G. Meikle (Ed.), *The Routledge companion to media and activism* (pp. 367-375). Routledge.

Treré, E. (2019). *Hybrid media activism: Ecologies, imaginaries, algorithms.* Routledge.

Tufekci, Z. (2017). *Twitter and tear gas: The power and fragility of networked protest.* Yale University Press.

Van Dijck, J. (2014). Datafication, dataism and dataveillance: Big data between scientific paradigm and ideology. *Surveillance & Society, 12*(2), 197–208. https://doi.org/10.24908/ss.v12i2.4776

Van Dijck, J., Poell, T., & De Waal, M. (2018). *The platform society: Public values in a connective world.* Oxford University Press.

Velkova, J., & Kaun, A. (2021). Algorithmic resistance: Media practices and the politics of repair. *Information, Communication & Society, 24*(4), 523–540. https://doi.org/10.1080/1369118X.2019.1657162

Willson, M. (2017). Algorithms (and the) everyday. *Information, Communication & Society, 20*(1), 137–150. https://doi.org/10.1080/1369118X.2016.1200645

Woolley, S. C., & Tucker, J. A. (2020). Bots and computational propaganda; Automation for communication and control. In N. Persily (Ed.), *Social media and democracy* (pp. 89–110). Cambridge University Press.

Yang, G. (2016). Narrative agency in hashtag activism: The case of #blacklivesmatter. *Media and Communication, 4*(4), 13. https://doi.org/10.17645/mac.v4i4.692

Zuboff, S. (2019). *The age of surveillance capitalism: The fight for a human future at the new frontier of power.* Profile Books.

# When digital capitalism takes (on) the neighbourhood: data activism meets place-based collective action

Vassilis Charitsis and Mikko Laamanen

**ABSTRACT**
Recent social movement scholarship has highlighted the instrumental and material roles played by digital technologies in supporting collective action and new cultures of organizing. This research seldom considers the symbolic role of digital technologies and data. However, the extraction and exploitation of data, described as data colonialism, facilitate novel opportunities for capitalist expansion in the everyday. Extensive data appropriation and commodification have led to a growing interest in data activism which challenges dominant data politics. Focusing on the intersection of data activism and local organising of collective action, this article examines two cases of how Big Tech disrupts everyday life and becomes a grievance used for mobilisation. With these cases, we illustrate how protest campaigns react to Google's aim to colonise both digital and physical spheres of life. The first case concerns the creation of a Google Campus in Berlin, while the second focuses on the Sidewalk Toronto urban development project led by the Google subsidiary, SideWalk Labs. Both projects were met with resistance comprising elements of data activism, mobilised as the Fuck Off Google and #blocksidewalk campaigns. Beyond rallying local discontent around impending gentrification and increased housing prices, the campaigns raised awareness about digital giants' unethical data practices and underscored alternative human-centric technologies and data politics. Employing frame analysis, we elaborate on the intersecting dynamics of traditional, community-based grassroots mobilisation and data activism against Googlization and explore the potentialities and limitations toward contextualising collective action *for* (rather than in) the digital age.

## Introduction

The unhampered dominance of global digital monopolies, the algorithmic manipulation of individual and public sentiments, the constant surveillance of users' activities and the appropriation and commodification of their data are some of the main reasons why the internet is increasingly perceived as "broken" (Berners-Lee, 2017; Jungherr et al., 2020; Streitfeld, 2017). The early celebratory discourse concerning the emancipatory potential of worldwide connectivity of information and people has given way to a sombre realisation that the commercial development of the internet has failed to give rise to a more democratic, inclusive and egalitarian world (Foster & McChesney, 2011). Digital capitalism has instead opened wide-ranging opportunities for the expansion and generalisation of capitalist extraction and exploitation (Schiller, 1999), particularly for those market actors that are able to develop monopolistic market positions and infrastructures (van Dijck et al., 2018). Indeed, wealth and power have become increasingly concentrated in a small number of Big Tech powerhouses: Google, Apple, Facebook, Amazon and Microsoft (or GAFAM) aim to colonise and commodify everyday life through data appropriation and dispossession (Couldry & Mejias, 2019; Thatcher et al., 2016). Vaidhyanathan's (2011) description of the 'Googlization of everything' captures precisely Big Tech's colonial ambitions, which extend to all spheres of life[1].

Digital capitalism has become a contentious issue at the level of data. As ownership and control of data becomes problematised, there is growing interest in data activism, which explores ways of resisting corporate appropriation of user data and opens the horizon for a citizen-centric data economy (Kennedy, 2018; Lehtiniemi & Ruckenstein, 2019). In social movement studies to date, digital technology is considered mainly in its instrumental capacity: namely how digital tools support activist mobilisation and new cultures of organizing. Digital technology also has material existence in software and hardware, infrastructures and skills in use. Importantly, yet to a limited extent, digital technologies are discussed symbolically in terms of their place in society and their impact on the kinds of futures they create and maintain (Mattoni, 2013). In data activism scholarship, Beraldo and Milan (2019) illustrate similar instrumental and symbolic dimensions of data, manifested in data-enabled and -oriented activism. On the one hand, data can provide significant support and thus enable both individual and collective action. On the other hand, data *itself* can become the locus of political contestation. Yet, the intangibility of data obfuscates the object of political action and hampers data-oriented mobilisations.

The expansion of digital capitalism is nevertheless not limited to opaque datafication processes but is manifested in everyday life in very concrete ways. Smartification and platformisation have transformed everyday life in urban environments towards complete control: this is emblematic of digital capitalism, where platforms serve to enclose the digital sphere and extract value from all activities (Sadowski, 2020). The rise of smart and platform urbanism presents an example of digital capitalism's desire to command and colonise everyday life. Reifying neoliberal ideals, the smart city bows to the hegemony of the market as the optimum, and uncontested, organisational mechanism (Morozov & Bria, 2018), which recasts citizens as consumers whose individual well-being is dependent on their financial, social, cultural and, increasingly, their data capital (De Waal & Dignum, 2017; Kitchin, 2019; Sadowski, 2019). Despite claims for enhanced civic engagement, security and convenience, corporate-led smart/platform urbanism reduces citizenship to data provision, enabling the privatisation of public spaces and services and the commodification of private life (Artyushina, 2020; Grossi & Pianezzi, 2017). Conforming to dominant data politics of hyper-individualism (Ruppert et al., 2017), smart urbanism exacerbates rather than addresses economic inequalities and social injustices (Cugurullo, 2018; Wiig, 2016). Imbued with free market ideals, data-driven urbanism legitimises race, gender and class discrimination (Benjamin, 2019; Eubanks, 2018; Noble, 2018), exerting algorithmic violence that further marginalises minorities and penalises underprivileged communities (Safransky, 2020). For instance, studies have shown that data-driven policing perpetuates historical biases and injustices against communities of colour, foreclosing their future (Ferguson, 2017; Richardson et al., 2019). Likewise, platformisation obscures exploitation, economic asymmetries, and physical and digital dispossessions (Schor, 2020; Slee, 2015). Lax regulation means that not only can digital companies avoid taxes and exploit their workforce (Laamanen et al., 2018), but colonise and commodify the data that workers and users generate.

Experience of such negative effects in everyday life illuminates the adverse consequences of data colonialism, which provide an impetus for citizen mobilisation. In this article, focusing on two cases that highlight Google's efforts to colonise the digital and physical spheres of life, we elaborate on the convergence of data activism with grassroots mobilisation to challenge hegemonic digital capitalism. Our first case concerns the creation of a Google Campus in Berlin, while the second case focuses on the Sidewalk Toronto urban development project led by the Google affiliate, SideWalk Labs. Both projects were met with digital counter-mobilisation combined with community-organised grassroots protest in the Fuck Off Google and #blocksidewalk campaigns, respectively. These campaigns framed their claims beyond the local implications to liveability to raising awareness concerning Google's (and Big Tech's) unethical data (and corporate) practices. Beyond a diagnosis of immediate problems to particular neighbourhoods (e.g. impending commercialisation and gentrification), the campaigns worked towards illustrating alternative human-centric technologies and data politics.

Our contribution acknowledges that the digital world cannot be free from existing social inequalities and injustices (Selwyn, 2004). Still, concerns about data-driven surveillance tend to be disconnected from other social justice issues (Dencik et al., 2016). Addressing these issues, we advance two key points. First, we develop the embryonic scholarship on data activism by highlighting the need to situate data-oriented activism within a broader framework of socio-political activism and link it to local struggles and activism. Second, we look beyond the instrumental role of digital technologies as movement-organising tools by drawing attention to collective action that challenges global data capitalism. Our analysis of the symbolic role of data develops a proposal of how local openings of resistance create opportunities to develop a data activist agenda on a global scale. With this, we complement extant analyses of social movements in the digital age with an argument towards studies of social movements *for* the digital age. This distinction foregrounds the need to pay attention to emergent collective action that addresses precisely the inequalities created and reinforced by digital capitalism and aims to lay the foundations for a just digital future, locally and translocally.

The article is structured as follows. We first discuss the role of data in digital capitalism and examine emerging critical scholarship on data activism. We incorporate concurrent social movement theorisation on the affordances provided by digital technologies in collective action but explore the emergence of mobilisation against the capitalism-driven disruption of everyday life and potential openings for the politicisation of data in two campaigns: Fuck Off Google in Berlin and #blocksidewalk in Toronto. Based on these cases, we conclude on the potentialities and limitations of place-based data activism against global data monopolists and identify pertinent implications and future directions for research intersecting data activism and social movement scholarship.

## Digital capitalism, data colonialism and data activism

Data has become the most important asset in the digital economy. Srnicek (2017, p. 6) explicates how data has been used as a means of revitalizing global capitalism: '... to maintain economic growth and vitality in the face of a sluggish production sector ... data have become increasingly central to firms and their relations with workers, customers, and other capitalists'. With the surge of data capitalism, data should not be ignored in any analysis of contemporary capitalism (Fuchs & Chandler, 2019; Morozov, 2015). As early as 2000, Haggerty and Ericson claimed that the production of surplus value is not limited to traditional wage labour relations but increasingly is the outcome of constantly monitored cybernetic activities. Terranova (2000) further argued that these activities constitute a form of free labour, with the digital economy being built on unwaged, exploited yet enjoyable user labour. Possibilities for endless surplus value extraction open up as labour time extends into leisure time, with leisure activities becoming appropriated and subsumed by capital (Prodnik, 2012). The diverse and pervasive surveillance mechanisms lead to the quantification and capture of personal and social life (Zuboff, 2019).

Data capture enables incessant commodification, with exploitation extending beyond the working day (Fuchs, 2014). Capital accumulation is no longer restricted to the exploitation of productive labour but also engulfs reproductive facets of private and social life, such as knowledge, health, leisure or culture (Lazzarato, 2004). When subjectivities, communication, creativity, feelings or thinking are captured and valorised by capital, we experience the real subsumption of society by capital (Read, 2009). This is the consequence of a new form of colonialism: data colonialism (Couldry & Mejias, 2019; Thatcher et al., 2016). The relentless datafication of everyday life not only colonises and commodifies spheres of life previously untouched by capital(ism) (Thatcher et al., 2016) but also reconfigures people into constant producers of value through data generation (Charitsis, 2016). Thus, data colonialism ushers in a new stage in capitalist production where human existence itself becomes a primary target for capital extraction and accumulation (Couldry & Mejias, 2019) by a few digital powerhouses, which strive to become the operating system of our lives (Vaidhyanathan, 2018). Data colonialism rests on pervasive surveillance – surveillance realism – which, with its integral and seemingly inescapable nature in the personal and social life, creates feelings of resignation against its domination (Dencik & Cable, 2017).

Refusing to accept the indomitability of data colonialism, the emergence of data justice and data activism explores ways to combat the inevitability of surveillance realism and develop the foundations for a just and equitable data and digital economy. Taylor (2017, p. 1) explicates that data justice refers to 'fairness in the way people are made visible, represented and treated as a result of their production of digital data' and maintains that data justice needs to be based on three pillars: (in)visibility, (dis)engagement with technology, and antidiscrimination. Achieving data justice requires a multifaceted confrontation of established data politics and practices of data capture. Data politics encapsulate political struggles pertaining to data collection and utilisation alongside the novel power relations that come into play in datafied societies (Ruppert et al., 2017). Emerging scholarship on data activism attempts to explore and analyse precisely the political struggles that address the issues of data capture and appropriation. Beraldo and Milan

(2019, p. 1) delineate the contentious politics of data as 'the bottom-up, transformative initiatives interfering with and/or hijacking dominant processes of datafication, contesting existing power relations or reappropriating data practices and infrastructure for purposes distinct from the intended'.

Early analyses indicated that data activism comprises both reactive and proactive engagement. Reactive engagement refers to solutions, mainly technological, that aim to fight against the ill-effects of massive data capture and accumulation, while proactive engagement explores the opportunities offered by datafication for civic engagement and social change (Beraldo & Milan, 2019; Milan & Gutiérrez, 2015). Advancing this analysis, Beraldo and Milan (2019) identify two roles for data in activism: an instrumental role as modular tools, or *data as repertoires*, and data as a political object of contestation in its own right, or *data as stakes*. For open data activists, datafication enhances their agency, as it empowers them and allows them to consider and develop alternative ways of collecting, using and thinking about data (Baack, 2015). However, the question of whether technological solutions alone can provide solid and adequate responses becomes critical when taking into consideration that Big Tech's colonial project extends to all spheres of life.

Dissenting voices and technologies do not escape the reach of data colonialism: on the contrary, they risk becoming co-opted and subsumed into the colonial logic. This remains the limitation of techno-solutionist, one-dimensional responses to data colonialism. Examining the development of data activism, Lehtiniemi and Ruckenstein (2019) identified the tensions between opposing social imaginaries prevalent in data activism: a technological imaginary, focussing on providing technological answers and responses; and a socio-critical imaginary, highlighting the political and ideological underpinnings of data conflicts. The latter stresses the inefficiency of technological corrections and acknowledges that technology alone will not resolve data injustices and power imbalances. Advancing a socio-economic critique, Couldry and Mejias (2019) maintain that resistance to the colonisation and appropriation of life can only take place if data colonialism is opposed at its central fundamental premise: omnipresent appropriation of data should not be seen as a naturally occurring process of social knowledge but as a commercially driven one that promotes specific economic interests to the detriment of human development.

In a similar vein, the digital as technologies and infrastructures appears in studies of social movements and collective action. The main perspective is instrumental, with a focus on organisational hybridity afforded by the instrumental, communicative and connective dimensions of digital technology (Earl, 2019; Funke & Wolfson, 2014). Digital technologies are seen to render collective action less cost-intensive (Bennett & Segerberg, 2012; Earl & Kimport, 2011) and to allow the emergence and sustenance of large-scale and even transnational mobilisation, such as the World Social Forum, the Arab Spring, or the Occupation of the Squares (Gerbaudo, 2012; Vicari, 2014). Technologies also have materiality in the hardware/software and infrastructure used in initiatives on a daily basis and between events (Costanza-Chock, 2012; Mattoni, 2013) In the material sense, technological access, knowledge, and competences have raised initial questions (since refuted: see Elliott & Earl, 2018) regarding inequalities of engagement in digital protest when this requires technological capital (see also Selwyn, 2004). While this scholarship underscores the socio-political dimensions of data activism, little attention has been paid to collective mobilisation against the ill effects of digital capitalism itself in everyday life.

The fact that a breakdown of everyday life can trigger collective action is the central idea of the quotidian disruption model in social movement theory (Borland, 2013). The breakdown of normal practices, spaces, and rhythms renders everyday life fragile, opens cleavages between societal actors, and importantly, establishes grievances that mobilise collective responses (Forno & Wahlen, 2022). In the context of neoliberal globalisation that centralises power to few central economies and large multinational corporations, resistance has relocalised to ward off the extraction of value from local communities and to create politics that are responsive to the community (Gibson-Graham, 2006). Urban movements aim to create self-sufficiency and governance around housing and utilities, health care, education and food provisioning, and even planning of the cityscape (della Porta & Diani, 2020; Laamanen & Wahlen, 2019). Place-based mobilisations allow for people to organise, to defend against and to attack the hypermobile, transnational, and overall rootless capitalism (Merrifield, 2003). Data colonialism encapsulates data capture, intensifying citizen surveillance, amplifying (labour) exploitation, and impairing the liveability and affordability of cities and neighbourhoods.

In this manner, the ill effects of digital capitalism engender sentiments of disenchantment among the affected communities. These sentiments can mobilise communicative action where activists purposefully manipulate and condense political, cultural and economic discourses to impact internal and external behaviours, boundaries and identities. Framing is a practice in the politics of signification (Snow & Benford, 1988) in relation to objects, situations, events and experiences of the participants, opponents and audiences. Claims, shared meaning, and solutions around a common grievance create crucial commitment and weld the participants together. With regard to the contentious politics of data (Beraldo & Milan, 2019), framing practices can illuminate the ways in which digital capitalism disrupts the everyday with data commodification and value extraction as well as cultural gentrification. As a contentious symbolic issue, data can be politicised when framed as a grand narrative of socio-economic re-engineering and exploitation or how the global networked and platform-based digital economy transforms society. While data capitalism has led to, or at least amplified, disruptions of the everyday, it has also expanded and normalised hyper-individualism, privileging atomised responses (Lake, 2017), which maintains the resilience of the system. Breaking away from this individualistic ethos, the two cases that are presented in the following section foreground the need for collective action and local mobilisation.

## Resisting googlization: a study of two campaigns

Google is at the forefront of data colonialism, given its aspirations not only to dominate the digital landscape but also to penetrate and colonise all aspects of life (Vaidhyanathan, 2011). Resisting the Googlization of life, even if desirable, appears increasingly futile. However, as Fisher (2009) underlines, the ubiquitous hegemony of capitalist realism should not be treated with fatalistic apathy: its perceived indomitability ultimately leads even small cracks on its edifice to open a horizon of alternatives. Similarly, small acts of avoidance and resistance to pervasive data capture in everyday life can play an important role in proving that data capitalism is not an inescapable reality. This brings us to the presentation of two cases of contentious data politics, the Fuck Off Google and

#blocksidewalk campaigns, contextualised and linked to community-based social issues in which data activism developed close connections with other forms of activism.

These are illustrative cases of emerging framing of data activist mobilisation that contests the physical and digital colonisation of everyday life. To elaborate the mobilising narratives of these mobilisations, we studied the online material made available by the companies involved in the proposed projects and by the two campaigns against them[2]. We also reviewed newspaper reporting on the mobilisations. We utilise a frame analysis framework (Benford & Snow, 2000; Snow & Benford, 1988) whereby activists are tasked to construct convincing (1) diagnostic framing that problematises a troublesome issue and focuses blame on an opponent; (2) prognostic framing entailing an elaboration of the remedial strategy towards a solution (or at least some tactical action towards this goal); and (3) motivational framing to incite action in participants. To mobilise support and action in participants, frames need to strike a chord with their cognitive schemata, both locally and globally.

### *Fuck Off Google (Berlin)*

In 2016, Google expressed interest in creating a Google Campus in the Umspannwerk power station in the Kreuzberg district of Berlin, Germany. The Google Campus concept has been established in several other locations globally as an incubator for start-ups in the digital business sector. Given that Kreuzberg is considered a counterculture hotbed and a bastion of anarcho-leftist activism (Gerbaudo, 2017), it is not surprising that a protest movement calling itself Fuck Off Google emerged in the neighbourhood.

> Google steals and exploits our data for profit and turned this behaviour into a norm. It also colonizes our physical spaces . . . As a decentralized network of people, we want to keep our lives and spaces free from this law- and tax-evading company and its peers and oppose the dystopian future they offer. (https://fuckoffgoogle.de)

Data was a central starting point for resistance here. Nevertheless, the initial campaign against the Google campus arose from a tangible threat to the neighbourhood. The Umspannwerk is located next to the Landwehrkanal in the Kreuzkölln area (a portmanteau of Kreuzberg and the neighbouring district of Neukölln), which has been strongly gentrifying since the 2010s and where protest is already on the daily agenda. The foundation was already laid for a multifaceted online and offline mobilisation. Online, the Fuck Off Google mobilisation grew with the fuckoffgoogle.de website, where the rationale for opposing the Google campus in Berlin is provided. Google is "G-entrificating" all, and will add to the threat of rising rents and diminishing liveability of the neighbourhood and its businesses. The disruption goes deeper, with the Google Campus seen to extract local skills, ideas and data from Berlin for the profit of Google, making the initiative more a farm than a campus. The website is complemented with a wiki page where users can contribute further information and material about the issues presented on the website. One of the main functions of the wiki page is to inform on upcoming events and provide information about physical protest action and materials, such as posters and flyers to be posted around the neighbourhood. The offline mobilisation consisted primarily of two types of activities: the monthly "noise events" where people gathered around the site of the proposed Google campus and engaged in loud protests each Friday, and the bimonthly "face2face café" meetings to discuss campaign issues and tactics. A newspaper entitled Shitstorm was published and street art produced (and documented on the wiki page).

The protest campaign achieved success when in late 2018 Google announced that it would abandon its original plans for the campus. Instead, Google offered the space rent-free to locally active non-profit organisations. Although the Fuck Off Google campaign welcomed the news, the campaigners cautioned about Google's ethically questionable whitewashing effort. As the activists elaborated:

> Karuna [one of the non-profits using the Umspannwerk space] publishes the homeless newspaper ... [which] is purely an advertising and propaganda magazine. In issue #5 this can be recognized by the design alone, for which the well-known Google colors were used ... It is especially ironic to see that homeless people are now walking around the city distributing a publication that contains content opposing their own interests. In short, Google indirectly uses homeless people for corporate propaganda. How perfidious and disgraceful is that? (https://fuckoffgoogle.de/g-entrificating-all/)

Importantly, Google retains the rights to the space and can resume its original plans in the future. As the front page of the Fuck Off Google campaign's website claims, the battle may be won but the fight is not over: 'Google canceled its plan for a "Google Campus" in Kreuzberg, Berlin under pressure from the neighborhood; yet it continues its expansion worldwide'. Beyond the local mobilisation against Google's plans for Kreuzberg, the Fuck Off Google wiki provided additional resources relating to Google's more general negative effects. Thus, the campaign was not only about the local space but also about people's everyday lives globally. The wiki has kept running ' ... to collect and aggregate data, campaigning material, and everything that can help kicking the "Google Campus" out of Kreuzberg, and Google out of our lives!' Thus, activists aim to engage further community action by highlighting Google's universal ambitions. In this manner, the campaign has developed a more comprehensive attack on Google's position and impact on societies, in addition to the harmful outcomes that the Google campus would have on the local area. Under such headings as "Google is evil" and "A brave G-world", it explores Google's role in mass surveillance, its involvement in the US drone program, its ambitions to dominate diverse fields of life (including healthcare, education, and entertainment), and the dystopian vision for smart cities. It also documents Google's unethical tax- and law-evading practices and its excessive lobbying to undermine democratic processes.

Beyond this diagnostic framing, de-googlification – the use of alternative technological solutions to Google – is offered as a potential way out (i.e. a prognostic frame) of the ills of the current situation. Notably, such 'solution' does not present an anti-technological rhetoric but aims to highlight the need for decentralised technological infrastructures for the common good:

> We want a world where our communications and data is in our own hands, where technology can be trusted and under our control, to enable us to organize a fairer future for everyone. We want to build, organize and share the practice of such technology based on the principles of decentralized services, free/libre software and end-to-end encryption. (http://fuckoffgoogle.de/decentralize-all/)

The campaign wiki offers a comprehensive list of ethically conscious privacy protecting alternatives, and also provides information about like-minded movements, campaigns and initiatives across the world. It dedicates a special section to Google's Sidewalk Labs subsidiary and its Toronto smart city plans, with an open letter of solidarity to the #blocksidewalk campaign. This states that:

> ... in Berlin we demonstrated that through decentralized efforts, deep local solidarity and relentless sharing of knowledge, entire neighborhoods can become strong enough to defeat – even if symbolically – the strongest of colossus ... We strongly believe that Toronto can win against Alphabet/Sidewalk Labs and their plan to take over our lives and spaces! (https://wiki.fuckoffgoogle.de/index.php?title=Smart_City_by_Google_(Toronto,_Sidewalk_Labs))

### *#blocksidewalk (Toronto)*

Sidewalk Labs, a subsidiary of Google's parent company Alphabet, provides technological solutions to urban infrastructures. In 2016, Sidewalk Labs CEO Dan Doctoroff outlined the company's vision to revolutionise urban life by planning the cities of tomorrow "from the internet up". Sidewalk Labs' first big project was Sidewalk Toronto, which aimed at turning a previously neglected Toronto neighbourhood into a model smart district by 'advanc[ing] a new model of inclusive urban development along Toronto's eastern waterfront, striving for the highest levels of sustainability, economic opportunity, housing affordability, and new mobility' (http://sidewalklabs.com/toronto). The technological utopianism upon which Sidewalk Toronto was premised associated superior quality of life with ubiquitous connectivity and data exploitation (see Doctoroff, 2016). However, a number of local people perceived this as a dystopian urban reality of pervasive surveillance and corporate control. A group of concerned residents, including academics, technologists, data scientists, community organisers, policy strategists, and open government advocates, organised a campaign under the moniker #blocksidewalk against the proposed Google-affiliated project.

The campaign was launched in 2019 with the aim of shutting down the project. Its demand was to 'develop Toronto's waterfront for the benefit of Torontonians, not corporate shareholders' (http://blocksidewalk.ca/join). Foregrounding public interests over private ones was the main driving force of the campaign, explicitly outlined in the campaign values:

> Urban planning is something that happens between Torontonians and the City, focused on the public interest. Development should benefit the people of Toronto and should respond to the needs of Torontonians as expressed by Torontonians. Development should prioritize city needs first, not the needs and interests of a private corporation. (http://blocksidewalk.ca/about)

Underscoring lack of transparency, the campaign raised concerns about the increase in the cost of living and warned that the financial terms of the deal undervalued the waterfront land. It rebutted claims regarding the development of affordable housing and the creation of tech jobs. The grievances about gentrification on corporate captured public land were complemented by strong opposition towards Toronto Sidewalk's vision of a data-driven city. The campaign particularly emphasised the lack of citizen data protection and the erosion of privacy. Framing the project as a surveillance testbed, lack of consent and control over data gathering and use emerged as the most contentious issue. The campaign stated:

> There is no option for residents, workers or visitors to opt out of urban data collection, and no safeguards for children, but there should be. All Torontonians deserve the right to say no to ubiquitous surveillance. (http://blocksidewalk.ca/faq2)

The campaign expressed data concerns regarding algorithmic biases against underprivileged community segments exacerbating discrimination and marginalisation and Sidewalk Labs' data aspirations to reach the healthcare sector posing serious threats to public universal healthcare. The campaign mobilised to inform the public about the proposed plan and to develop alternative proposals for the rejuvenation of the city's waterfront. Public meetings and workshops were organised to discuss Sidewalk Toronto's potential negative effects on the city and public life. Campaign activists attended and lobbied against the project at public consultations organised by Waterfront Toronto (the public corporation set up in 2001 by the national and regional governments and the City of Toronto to oversee the waterfront revitalisation). Activists also joined events organised by Sidewalk Labs to voice their opposition. By organising press conferences and issuing a number of media releases, the campaign managed to attract considerable local, national and international media attention.

Beyond the immediate local threat, the campaign's website discusses in detail Google's record on breaching data protection laws, engaging in dubious deals to develop data centres, the treatment of employees, and investigations about antitrust violations. The campaign connected with other community-based organisations and drew support for many liked-minded international initiatives like RethinkLink (New York), Silicon Valley Rising (San Jose), Digital Commonwealth (Portland) and Fuck Off Google (Berlin). Ultimately, in May 2020, Sidewalk Labs withdrew from the project, citing financial uncertainty due to the ongoing global pandemic. Undeniably, the pressure from the campaign proved crucial in halting the Sidewalk Toronto project. A campaign activist underlined that the success sends 'a message to Silicon Valley on behalf of all those around the world who are fighting Big Tech in their cities' (http://blocksidewalk.ca/victory). While the main objective of the campaign was reached, the organisers emphasise the need for sustained organizing to prevent private technology companies from usurping control of public spaces, services and resources.

## Discussion

In a neoliberal society, economic actors and entities become targets of collective action in addition to, and even instead of, the more traditional political enemies. It is well-established that social movement action increasingly targets actors by manipulating economic and cultural narratives rather than political structures (de Bakker et al., 2017; della Porta, 2015). While capitalism has "gone digital", movements confronting data colonialism are yet to be meaningfully studied. This is surprising given that Big Tech are implicated in exercising power that disrupts the collective good in societies: they engage in tax-avoidance, rig competition, and undermine healthy labour relations; they build infrastructure to drive gentrification while simultaneously invading privacy through datafication practices and data capture that allow the commercialisation of the everyday.

Big Tech not only drives change in the use of technology, but also transforms the economy, democracy, and human relations (Galloway, 2017). Datafication is not only engrained in opaque technological processes, but also embedded within socio-economic and political contexts, both locally and internationally. At the local level, Big Tech domination is manifested through rampant gentrification, rising housing prices and increased homelessness, as exemplified by the socio-economic dynamics of Silicon Valley. Additionally, pervasive surveillance and the privatisation of public space are enclosing everyday life with data appropriation in a holistic, yet opaque way. This is amplified in the manner in which data-driven platforms serve as advertising media markets where users are micro-targeted through pervasive surveillance (Zuboff, 2019). Disturbingly, in these commercial spaces, data can be fed back to the users to confirm or manipulate their perception of the social world (Jungherr et al., 2020). Thus, data is a major source of corporate profit *and* societal disruption. Yet, responses to intrusive datafication tend to be one-dimensional, instrumental-technological solutions de-emphasizing broader socio-political forces, processes and implications. Similarly, it seems that data appropriation breeds resignation rather than resistance.

Therefore we underscore the symbolic role of data for collective action, acknowledging the multifaceted dimensions of big-tech global domination, which further highlight the limits of single-axis forms of resistance. Concurrently, the two cases outlined here demonstrate the importance of localised forms of resistance, as the Fuck Off Google and #blocksidewalk campaigns managed to strike small but important blows against the purported indomitability of digital capitalism. The framings of both campaigns oscillated between the local and the global context. On the one hand, the local character of the campaigns mobilised concrete grievances: this enabled the collective action frames to resonate with the communities in question. The diagnostic frame specified the opponent, Google, and the immediate threats to the local community in gentrification, privatisation and surveillance. The prognostic frame concentrated on preventing the proposed projects from materialising and advanced proposals for alternative digital and urban futures. Finally, the motivational framing called for active participation in action to fend off the opponent: 'Don't let Google & co. take over our lives and spaces!' (Fuck Off Google) and 'It's time to take our city back' (#blocksidewalk).

On the other hand, the translocal character across the two campaigns focused on the ill effects of digital capitalism in general, and Google in particular, and provided (well-established) argumentation to how Googlization threatens individuals and communities. Translocal diagnostic frames further provided evidence of similar projects across the world, demonstrating negative impact on local communities. Instrumentally, technological developments allow for translocal communication and organising strategies from which local activists could draw and learn within the broader global network (Sterpka, 2013; Tsutsui & Smith, 2019). For McFarlane (2009, p. 561), such networks are *assemblages* of 'materially heterogeneous, practice-based, emergent and processual' conceptions of space, rather than mere connective (i.e. instrumental) networks. Translocal assemblages of protest are perhaps best illustrated in the efforts to mobilise the alter-globalisation movement. The mobilisation of this movement illustrated disdain for neoliberal capitalism and imperialism, its central actors and institutions (the WTO, IMF, WB and G8), and ways of resisting both in transnational solidarity networks. Effectively, much of the resistance took place in direct local confrontation and prefigurative organising (Laamanen, 2022) where digital technologies amplified collective identity and cross-national solidarity with the use of master frames that recognise the symbolic grievances across localities diffused across sites and repertoires of protest (Sádaba Rodríguez, 2020; Vicari, 2014).

Proximity remains a crucial element for maintaining long-term relations and active participation in digital movements (van Haperen et al., 2018). Protest that transcends geopolitical boundaries while at the same time remaining locally relevant in confronting neoliberal capitalism is illustrative of the place-based globalism that connects micro-political, place-based processes across geographies (Osterweil, 2005). The cases in this article illustrate the importance of place-based mobilisation, where local resistance creates the primary spaces for movement claims making with potential spill-over of these claims to mobilise other localities experiencing (similar) disruptive conditions. Motivational framing highlighted links and solidarity ties with like-minded protest campaigns across the world: 'From Berlin to Toronto, to Queens, to SF, San Jose or Rennes, you can count on our undivided love, attention and solidarity!' (https://wiki.fuckoffgoogle.de/index.php?title=Smart_City_by_Google_(Toronto,_Sidewalk_Labs)).

Overall, prognostic translocal frames are less prominent. Yet, as the protest campaigns highlight, forms of data colonialist domination and authority overlap and intersect in urban spaces. The campaigns examined in this article acknowledged shared values and aims with similar campaigns across the world, expressing solidarity and support. In that respect, protesting digital capitalism can connect to the legacy of the alterglobalisation movement, creating connections, claims and practices that attempt to overcome disconnectedness and illustrate the global presence of a data countermovement. Further research is needed to explore whether and to what extent local campaigns against data colonialism create translocal links forming global networks of resistance and (trans)local solutions to fence out global data capitalists.

## Conclusion

Data's soaring value and demand render it a major issue of socio-political contention. Given the multidimensionality in the datafication and commercialisation of the everyday, protest campaigns need to escape the confines of single-axis frames and forms of resistance. Fuck Off Google and #blocksidewalk demonstrate that when linked to other forms of activism and embedded within a local context, data activism can provide successful responses to digital capitalism. The reterritorialisation of data activism enables collective action that can have a tangible impact at the local level and potentially erode data colonialism.

Complementing scholarship on the contentious politics of data (Beraldo & Milan, 2019), we examined data not as an isolated object of political struggle, but as one linked to specific place politics and social contexts. Place-based mobilisations against data colonialism constitute manifestations of emerging collective movements for the digital age. Defying surveillance realism, these movements address collective grievances against the dominance of digital capitalism and its disrupting consequences. The local articulations of data activism recognise that data politics and practices do not appear in a vacuum but are embedded in political, economic and social structures. At the same time, data politics engender specific social relations and living realities that disempower individuals and communities. Physical spaces become sites of contest as Big Tech's attempts to usurp and control them meets the collective resistance of local communities. Data as a contentious issue becomes entangled with the well-being of places and communities. For emergent social movements for the digital age, as indicated by the campaigns examined in this article, data politics are part of broader political projects. Contesting dominant politics goes hand-in-hand with challenging politics of urban inequalities, including those amplified by data colonialism.

What remains to be seen is whether such placed-based local initiatives can develop sustained and connected translocal forms of data activism that will powerfully challenge digital capitalist structures by providing convincing prognostic frames and alternatives. As Olesen (2015, p. 71) elaborates, 'the largely imagined nature of global grievance communities [necessitates] symbols that can serve as common reference and orientation points'. Attempts to confront Big Tech's global dominance require not only international mobilisation but also forging alliances and cooperation between users and tech workers (Larson, 2020). By focussing on the symbolic dimensions of technology, and particularly data, we are addressing a lacuna in current social movement research, which tends to focus on the instrumental (i.e. networking) and material (i.e. ITC resources and skills) dimensions (Mattoni, 2013) and overstate their importance (Flesher Fominaya, 2014). In that respect, this article has provided an opening for discussion regarding further analyses of social movements *for* the digital age. This highlights the necessity for further studies that will not only examine how digital technologies and data can support and maintain mobilisation but also explore further collective action in response to pervasive datafication processes and dominant data politics. As digital capitalism becomes an omnipresent commodifying force in people's lives through data capture, it is imperative to shed light on emergent collective action for a digital future that rejects Big Tech colonial visions and contests exploitative relations both locally and translocally.

## Note

1. Vaidhyanathan (2011) coined the phrase 'Googlization of everything' to illustrate Google's aim to expand into diverse areas of social life. Following Sharon (2016, 2021), we use the term googlization to refer not only to Google's colonial ambitions but also to those of Big Tech in general.
2. At the time of the data collection, all campaign webpages were live. At the time of publication, the blocksidewalk.ca pages can no longer be accessed while the campaign sites on Facebook (@BlockSidewalkTO) and Twitter (@blocksidewalk) can still be reached. In the data analysis, links to the campaign webpages are mentioned in full details.

## Acknowledgement

Early versions of the paper were presented at the 'Alternatives to Capitalism' research network meeting at the Society for the Advancement of Socio-Economics 2019 conference, the European Communication Research and Education Association 'Infrastructures and

Inequalities: Media Industries, Digital Cultures and Politics' 2019 conference and the Royal Holloway Digital Organisation and Society Research Centre 'Digital Activism' workshop in 2020.

## Disclosure statement

No potential conflict of interest was reported by the authors.

## Funding

This work was supported by the Royal Holloway University of London Digital Organisation and Society Research Centre ECR research grant.

## ORCID

Vassilis Charitsis http://orcid.org/0000-0003-3745-0120
Mikko Laamanen http://orcid.org/0000-0003-3737-9934

## References

Artyushina, A. (2020). Is civic data governance the key to democratic smart cities? The role of the urban data trust in Sidewalk Toronto. *Telematics and Informatics*, *55*, 101456. https://doi.org/10.1016/j.tele.2020.101456

Baack, S. (2015). Datafication and empowerment: How the open data movement re-articulates notions of democracy, participation, and journalism. *Big Data & Society*, *2*(2), 1–11. https://doi.org/10.1177/2053951715594634

Benford, R. D., & Snow, D. A. (2000). Framing processes and social movements: An overview and assessment. *Annual Review of Sociology*, *26*(1), 611–639. https://doi.org/10.1146/annurev.soc.26.1.611

Benjamin, R. (2019). *Race after technology: Abolitionist tools for the new jim code*. John Wiley & Sons.

Bennett, W. L., & Segerberg, A. (2012). The logic of connective action. Digital media and the personalization of contentious politics. *Information, Communication & Society*, *15*(5), 739–768. https://doi.org/10.1080/1369118X.2012.670661

Beraldo, D., & Milan, S. (2019). From data politics to the contentious politics of data. *Big Data & Society*, *6*(2), 1–11. https://doi.org/10.1177/2053951719885967

Berners Lee, T. (2017, March12). Three challenges for the web, according to its inventor. *World Wide Web Foundation*. https://webfoundation.org/2017/03/web-turns-28-letter

Borland, E. (2013). Quotidian disruption. In D. A. Snow, D. della Porta, B. Klandermans, & D. McAdam (Eds.), *The Wiley-Blackwell encyclopedia of social and political movements* . Wiley-Blackwell.

Charitsis, V. (2016). Prosuming (the) self. *Ephemera: Theory and Politics in Organization*, *16*(3), 37–59.

Costanza-Chock, S. (2012). Mic check! Media cultures and the occupy movement. *Social Movement Studies*, *11*(3–4), 375–385. https://doi.org/10.1080/14742837.2012.710746

Couldry, N., & Mejias, U. A. (2019). Data colonialism: Rethinking big data's relation to the contemporary subject. *Television & New Media*, *20*(4), 336–349. https://doi.org/10.1177/1527476418796632

Cugurullo, F. (2018). Exposing smart cities and eco-cities: Frankenstein urbanism and the sustainability challenges of the experimental city. *Environment and Planning A: Economy and Space, 50* (1), 73–92. https://doi.org/10.1177/0308518X17738535

de Bakker, F., den Hond, F., & Laamanen, M. (2017). Social movements: organizations and organizing. In C. Roggeband & B. Klandermans (Eds.), *Handbook of social movements across disciplines* (pp. 203–231). Springer.

della Porta, D. (2015). *Social movements in times of austerity*. Polity.

della Porta, D., & Diani, M. (2020). *Social movements: An Introduction* (3rd ed.). Wiley Blackwell.

Dencik, L., & Cable, J. (2017). The advent of surveillance realism: Public opinion and activist responses to the Snowden leaks. *International Journal of Communication, 11*, 763–781.

Dencik, L., Hintz, A., & Cable, J. (2016). Towards data justice? The ambiguity of anti-surveillance resistance in political activism. *Big Data & Society, 3*(2), 1–12. https://doi.org/10.1177/2053951716679678

De Waal, M., & Dignum, M. (2017). The citizen in the smart city. How the smartcity could transform citizenship. *It - Information Technology, 59*(6), 263–273. https://doi.org/10.1515/itit-2017-0012

Doctoroff, D. L. (2016, November30). Reimagining cities from the internet up. *Medium*. https://medium.com/sidewalk-talk/reimagining-cities-from-the-internet-up-5923d6be63ba

Earl, J. (2019). Technology and social media. In D. A. Snow, S. A. Soule, H. Kriesi, & H. J. McCammon (Eds.), *The Wiley Blackwell companion to social movements* (2nd ed., pp. 289–305). Wiley-Blackwell.

Earl, J., & Kimport, K. (2011). *Digitally enabled social change: Activism in the internet Age*. MIT Press.

Elliott, T., & Earl, J. (2018). Online protest participation and the digital divide: Modeling the effect of the digital divide on online petition-signing. *New Media & Society, 20*(2), 698–719. https://doi.org/10.1177/1461444816669159

Eubanks, V. (2018). *Automating inequality*. St. Martin's Press.

Ferguson, A. G. (2017). *The rise of big data policing: Surveillance, race, and the future of law enforcement*. New York University Press.

Fisher, M. (2009). *Capitalist realism: Is there no alternative?* Zero Books.

Flesher Fominaya, C. (2014). *SocIal movements and globalisation. how protests, occupations and uprisings are changing the world*. Red Globe Press.

Forno, F., & Wahlen, S. (2022). Prefiguration in everyday practices: When the mundane becomes political. In L. Monticelli (Ed.), *The future is now. An introduction to prefigurative politics* (pp. 119–129). Bristol University Press.

Foster, J. B., & McChesney, R. W. (2011). The internet's unholy marriage to capitalism. *Monthly Review, 62*(10), 1–30.

Fuchs, C. (2014). Digital prosumption labour on social media in the context of the capitalist regime of time. *Time & Society, 23*(1), 97–123. https://doi.org/10.1177/0961463X13502117

Fuchs, C., & Chandler, D. (2019). Introduction: Big data capitalism-politics, activism, and theory. In D. Chandler & C. Fuchs (Eds.), *Digital objects, digital subjects* (pp. 1–20). University of Westminster Press.

Funke, P. N., & Wolfson, T. (2014). Class in-formation: The intersection of old and new media in contemporary urban social movements. *Social Movement Studies, 13*(3), 349–364. https://doi.org/10.1080/14742837.2013.831755

Galloway, S. (2017). *The four: The hidden DNA of Amazon, Apple, Facebook, and Google*. Portfolio.

Gerbaudo, P. (2012). *Tweets and the streets: Social media and contemporary activism*. Pluto Press.

Gerbaudo, P. (2017). *The mask and the flag: Populism, citizenism and global protest*. Oxford University Press.

Gibson-Graham, J. K. (2006). *A postcapitalist politics*. University of Minnesota Press.

Grossi, G., & Pianezzi, D. (2017). Smart cities: Utopia or neoliberal ideology? *Cities, 69*, 79–85. https://doi.org/10.1016/j.cities.2017.07.012

Haggerty, K. D., & Ericson, R. V. (2000). The surveillant assemblage. *The British Journal of Sociology*, *51*(4), 605–622. https://doi.org/10.1080/00071310020015280

Jungherr, A., Rivero, G., & Gayo-Avello, D. (2020). *Retooling politics: How digital media are shaping democracy*. Cambridge University Press.

Kennedy, H. (2018). Living with data: Aligning data studies and data activism through a focus on everyday experiences of datafication. *Krisis: Journal for Contemporary Philosophy*, *1*, 18–30.

Kitchin, R. (2019). Toward a genuinely humanizing smart urbanism. In P. Cardullo, C. DiFeliciantonio, & R. Kitchin (Eds.), *The right to the smart city* (pp. 193–204). Emerald Publishing.

Laamanen, M. (2022). Organising prefiguration. In L. Monticelli (Ed.), *The future is now: An introduction to prefigurative politics* (pp. 191–203). Bristol University Press.

Laamanen, M., Barros, M., & Islam, G. (2018). Collective representation on collaborative economy platforms. In I. S. Cruz, R. Ganga, & S. Wahlen (Eds.), *Contemporary collaborative consumption* (pp. 35–56). Springer.

Laamanen, M., & Wahlen, S. (2019). Sharing economy and lifestyle movements. In R. Belk, G. M. Eckhardt, & F. Bardhi (Eds.), *Handbook of the sharing economy* (pp. 49–60). Edward Elgar.

Lake, R. (2017). Big Data, urban governance, and the ontological politics of hyper-individualism. *Big Data & Society*, *4*(1), 1–10. https://doi.org/10.1177/2053951716682537

Larson, R. (2020). *Bit tyrants: the political economy of silicon valley*. Haymarket.

Lazzarato, M. (2004). From capital-labour to capital-life. *Ephemera: Theory and Politics in Organization*, *4*(3), 187–208.

Lehtiniemi, T., & Ruckenstein, M. (2019). The social imaginaries of data activism. *Big Data & Society*, *6*(1), 1–12.

Mattoni, A. (2013). Technology and social movements. In D. A. Snow, D. della Porta, B. Klandermans, & D. McAdam (Eds.), *The Wiley-Blackwell encyclopedia of social and political movements*. Wiley-Blackwell.

McFarlane, C. (2009). Translocal assemblages: Space, power and social movements. *Geoforum*, *40* (4), 561–567. https://doi.org/10.1016/j.geoforum.2009.05.003

Merrifield, A. (2003). Death on credit. *Antipode*, *35*(1), 197–200. https://doi.org/10.1111/1467-8330.00314

Milan, S., & Gutiérrez, M. (2015). Citizens' media meets big data: The emergence of data activism. *Mediaciones*, *11*(14), 120–133. https://doi.org/10.26620/uniminuto.mediaciones.11.14.2015.120-133

Morozov, E. (2015, June23). Digital technologies and the future of data capitalism. *Social Europe*. https://www.socialeurope.eu/digital-technologies-and-the-future-of-data-capitalism

Morozov, E., & Bria, F. (2018). *Rethinking the smart city. Democratizing urban technology*. Rosa Luxemburg Stiftung.

Noble, S. U. (2018). *Algorithms of Oppression*. New York University Press.

Olesen, T. (2015). *Global injustice symbols and social movements*. Palgrave Macmillan.

Osterweil, M. (2005). Place-Based globalism: Theorizing the global justice movement. *Development*, *48*(2), 23–28. https://doi.org/10.1057/palgrave.development.1100132

Prodnik, J. (2012). A note on the ongoing processes of commodification: From the audience commodity to the social factory. *tripleC: Communication, Capitalism & Critique*, *10*(2), 274–301. https://doi.org/10.31269/triplec.v10i2.409

Read, J. (2009). A genealogy of homo-economicus: Neoliberalism and the production of subjectivity. *Foucault Studies*, *6*, 25–36. https://doi.org/10.22439/fs.v0i0.2465

Richardson, R., Schultz, J. M., & Crawford, K. (2019). Dirty data, bad predictions: How civil rights violations impact police data, predictive policing systems, and justice. *New York University Law Review*, *94*, 15–55.

Ruppert, E., Isin, E., & Bigo, D. (2017). Data politics. *Big Data & Society*, *4*(2), 1–7. https://doi.org/10.1177/2053951717717749

Sádaba Rodríguez, I. (2020). The technopolitical frameworks of contemporary social movements. In C. F. Fominaya & R. A. Feenstra (Eds.), *Routledge handbook of contemporary European social movements* (pp. 313–325). Routledge.

Sadowski, J. (2019). When data is capital: Datafication, accumulation, and extraction. *Big Data & Society*, *6*(1), 1–12. https://doi.org/10.1177/2053951718820549

Sadowski, J. (2020). *Too smart: How digital capitalism is extracting data, controlling our lives, and taking over the world.* MIT Press.

Safransky, S. (2020). Geographies of algorithmic violence: Redlining the smart city. *International Journal of Urban and Regional Research*, *44*(2), 200–218. https://doi.org/10.1111/1468-2427.12833

Schiller, D. (1999). *Digital Capitalism.* MIT Press.

Schor, J. (2020). *After the gig: How the sharing economy got hijacked and how to win it back.* University of California Press.

Selwyn, N. (2004). Reconsidering political and popular understandings of the digital divide. *New Media & Society*, *6*(3), 341–362. https://doi.org/10.1177/1461444804042519

Sharon, T. (2016). The Googlization of health research: From disruptive innovation to disruptive ethics. *Personalized Medicine*, *13*(6), 563–574. https://doi.org/10.2217/pme-2016-0057

Sharon, T. (2021). From hostile worlds to multiple spheres: Towards a normative pragmatics of justice for the Googlization of health. *Medicine, Health Care, and Philosophy*, *25*(3), 315–327. https://doi.org/10.1007/s11019-021-10006-7

Slee, T. (2015). *What's yours is mine: Against the sharing economy.* OR Books.

Snow, D. A., & Benford, R. D. (1988). Ideology, frame resonance, and participant mobilization. In B. Klandermans, H. Kriesi, & S. Tarrow (Eds.), *From structure to action: social movement participation across cultures: International social movement research* (Vol. 1, pp. 197–217). JAI Press.

Srnicek, N. (2017). *Platform Capitalism.* Polity.

Sterpka, M. K. (2013). The transnational struggle for information freedom. In J. S. Juris & A. Khasnabish (Eds.), *Insurgent encounters: transnational activism, ethnography, and the political* (pp. 295–317). Duke University Press.

Streitfeld, D. (2017, May20). 'The internet is broken': @ev is trying to salvage it. *New York Times.* https://www.nytimes.com/2017/05/20/technology/evan-williams-medium-twitter-internet.html

Taylor, L. (2017). What is data justice? The case for connecting digital rights and freedoms globally. *Big Data & Society*, *4*(2), 1–14. https://doi.org/10.1177/2053951717736335

Terranova, T. (2000). Free labor: Producing culture for the digital economy. *Social Text*, *18*(2), 33–58. https://doi.org/10.1215/01642472-18-2_63-33

Thatcher, J., O'Sullivan, D., & Mahmoudi, D. (2016). Data colonialism through accumulation by dispossession: New metaphors for daily data. *Environment and Planning D Society & Space*, *34* (6), 990–1006. https://doi.org/10.1177/0263775816633195

Tsutsui, K., & Smith, J. (2019). Human rights and social movements. In D. A. Snow, S. A. Soule, H. Kriesi, & H. J. McCammon (Eds.), *The Wiley Blackwell companion to social movements* (2nd ed., pp. 586–601). Wiley-Blackwell.

Vaidhyanathan, S. (2011). *The googlization of everything (and why we should worry).* University of California Press.

Vaidhyanathan, S. (2018). *Antisocial media: How Facebook disconnects us and undermines democracy.* Oxford University Press.

van Dijck, J., Poell, T., & de Waal, M. (2018). *The platform society: PubLic values in a connective world.* Oxford University Press.

van Haperen, S., Nicholls, W., & Uitermark, J. (2018). Building protest online: Engagement with the digitally networked #not1more protest campaign on Twitter. *Social Movement Studies*, *17* (4), 408–423. https://doi.org/10.1080/14742837.2018.1434499

Vicari, S. (2014). Networks of contention: The shape of online transnationalism in early twenty-first century social movement coalitions. *Social Movement Studies, 13*(1), 92–109. https://doi.org/10.1080/14742837.2013.832621

Wiig, A. (2016). The empty rhetoric of the smart city: From digital inclusion to economic promotion in Philadelphia. *Urban Geography*, *37*(4), 535–553. https://doi.org/10.1080/02723638.2015.1065686

Zuboff, S. (2019). *The age of surveillance capitalism: The fight for the future at the new frontier of power*. Profile Books.

OPEN ACCESS

# Doubt to be certain: epistemological ambiguity of data in the case of grassroots mapping of traffic accidents in Russia

Dmitry Muravyov

**ABSTRACT**

While the prevalent view positions data as an objective and unbiased resource of truth about the world, scholars have noted that this understanding cannot be all-encompassing and data activists may comprehend the relationship between knowledge, reality, and data differently. Data activists are civil society actors with a critical stance towards datafication; they either consider data as a political issue or employ it to advance desirable social change. This article investigates activists' data epistemologies in a twofold manner. First, it poses the question of how activists can simultaneously use a certain dataset while questioning its credibility. Second, the article explores how activists' data epistemology transforms other domains of socio-political grassroots interventions. To answer these questions, I turn to the case of the DTP Map – an interactive geoweb map of traffic accidents in Russia made by activists using the official governmental data. Turning to the concept of contentious data politics, I demonstrate how the project transforms by continuously dealing with the data's epistemologically ambiguous nature. In their data practices aimed at gaining and maintaining the users' trust, activists have tried to ensure their project will be employed by various collectives for the common goal of reducing traffic accidents in Russia. Their data practices can be considered both a repertoire of social change and a stake of activist intervention. Crucially, in the process of map-making, activists do not gain the epistemologically unambiguous view of data but rather they manage to retain this ambiguity and make it a constitutive part of their project.

## Introduction

A well-known saying in Russia goes like this: 'There are two troubles in Russia: fools and roads.' Humorous in nature, it nevertheless relates to the widespread and acknowledged infrastructural issues with roads in Russia and, by association, traffic accidents; alongside the state, urban activists are among those who try to address this issue. In this article, I examine one such activist project, the geoweb platform DTP Map, by studying how urban activists use government data on traffic accidents to advance social change in road infrastructure while simultaneously – and, seemingly, paradoxically – questioning the nature of this data.

The DTP Map (Karta DTP; Карта ДТП) is an interactive geoweb map of traffic accidents in Russia and an activist project whose aim is to draw attention to the issue of traffic accidents and provide journalists, urban activists, and researchers with an accessible platform to study it. Launched in 2018 by Sergey Ustinov, the lead developer, and Alexey Radchenko, the urban activist, the map relies on the official government statistics of the General Administration for Traffic Safety (GIBDD), a part of the Russian Ministry of Internal Affairs. The platform allows users to interactively examine data on traffic accidents for a certain time, filter it by participants, streets, types, weather, and many more thereby making possible more sophisticated forms of data analysis compared to the official representation of data on behalf of GIBDD (Figure 1). The GIBDD's data originally provoked activists' interest because it was abundant, detailed, and available. Moreover, that data has also not been previously used by other tech activists even though traffic accidents continue to remain a significant problem across Russia.

According to activists, the data has not been represented in a way that would engage people in seeing traffic accidents as an important urban and infrastructural problem. Yet it is not only about another way of representing the issue; what the DTP Map as a project has aimed to achieve is not only to rearrange how data are represented but also to do it in a manner that would generate the attention of urban activists, local journalists, and non-expert citizens.

The DTP Map as an instance of urban activism is enabled by datafication, which is the transformation of social phenomena or events (such as traffic accidents) into (online) quantified data (Van Dijck, 2014). Building on digital infrastructures and legacies of social statistics, datafication enacts new forms of both social structure and agency. Among the latter is data activism which is a concept that attempts to capture new forms of activism enabled by widespread datafication. Data activists both use and question social, political, and economic relations around data infrastructures to

**Figure 1.** The DTP Map Interface.

appropriate and subvert them to bring about the social change they consider desirable (Milan & Gutiérrez, 2015).

By considering the DTP Map as a case of data activism, it becomes possible to grasp how (urban) activism changes when it is re-shaped by datafication. Accordingly, the topic of data activism deserves attention from scholars of both social movement studies and critical data studies because the data activist collective and individual actions are conditioned by movements' structure, dynamics, and goals as well as on data politics; data activists rarely act alone, and their work happens in a certain socio-political context.

Following this line of interdisciplinary inquiry, Beraldo and Milan (2019) have advanced the notion of 'contentious data politics' to grasp how multiple forms of data-mediated collective action are exercised in the age of datafication. From the perspective of contentious data politics, the action of activists is re-mediated by data as the latter can be an object of activists' intervention (data as stake), for instance in the form of producing tools to protect users' data privacy, or as a repertoire of social change, e.g. when data are used as evidence of socio-political issues like gender or racial inequality (data as a repertoire).

Additionally, in the process of activist interventions, data are imagined by activists to provide a certain knowledge about the world. Milan and Van der Velden (2016) have suggested considering 'alternative epistemologies of data activism,' i.e. what kind of knowledge is produced through activists' data practices and how it differs from other 'data-driven' domains, such as business or government work.

If we agree that activists understand data differently, then from the theoretical perspective of contentious data politics, certain research questions emerge. How can activists doubt the data while advancing social change using it? How do these data-related epistemological ambiguities transform activists' data-based projects? Building on the detailed case study of the DTP Map, I employ the theoretical framework of contentious data politics to reflect on how concerns about data as a source of knowledge about the world become constitutive for the project's continuous development.

In this article, I shall focus on the epistemological ambiguity of the data in the DTP Map project as it is conditioned by the contentious dynamics of the data activist efforts situated between engaging data as either stakes or a repertoire (Beraldo & Milan, 2019) – I shall elaborate on that distinction in the next part of the article. I argue that data for the DTP Map project is not something pre-given and always exhaustively representative of the world; activists cannot take data for granted, and it continuously retains its epistemological ambiguity. The latter refers to a practical tension between complete trust in data (and, consequently, establishing a mimetic relationship between reality and data) and its total dismissal as a source of knowledge. Crucially, the key argument of the article is that epistemological ambiguity of data is a constitutive trait of the DTP Map project because it both ensures the possibility of using data as a repertoire of contention and allows to maintain the users' trust in the project.

My article aims to contribute to the interdisciplinary studies of data activism and social movements in three ways. First, it provides original research on the case of a data activist initiative in Russia; to this day, and to my knowledge, the study of data activism as a phenomenon has been scarcely documented in Russia. Second, as I show later in the article, the case of the DTP Map can be seen as a part of the broader ecology of urban social movement, which employs various strategies to change the urban infrastructure and landscape in Russia. Third, following Chesters's (2012) suggestion to consider social movement activism not purely as an object of academic research, but as a knowledge producer, I consider the alternative epistemologies of the DTP Map as a data activist initiative to elucidate how activists reflect on the epistemological ambiguity of data in their work and what sets of practices are enlisted in as part of the process.

This article is organised into two parts. In the first part, I elaborate on the theoretical framework of the article as well as introduce some information relevant to understanding data activism in Russia. In the second part, I provide a case study of the DTP Map to grasp how the ambiguous epistemological status of data is shaped by the activists' contentious practices.

## Theorising the DTP Map

In this part, I introduce the theoretical framework of the article. First, I will situate the scholarly concern about data activist epistemology within the discussions about social movements and knowledge, dataism, and alternative epistemologies. Second, I use the theoretical framework of 'contentious data politics' and briefly contextualise it within the Russian political and tech activist context. Third, I introduce the concept of epistemological ambiguity of data to elucidate how activists navigate in practice the challenge of employing data for social change while simultaneously distancing themselves from understanding it as a mimetic representation of an external reality.

### *Data, knowledge, and activism: between dataism and alternative epistemologies*

Broadly speaking, as many scholars have argued before, the multi-faceted proliferation of data and statistics in social life is accompanied by certain discourses, mythologies, and imaginaries that imbue them with the specific cultural meaning of rationality, efficiency, and objectivity (Desrosières, 1998; Hacking, 1990; Porter, 1986). What follows from this line of research is the need to comprehend the epistemological shifts brought by datafication as well as the political ramifications that different ways of understanding and enacting reality with data bring about.

Presently, with the rise of big data and algorithmic techniques, we are witnessing similar shifts in understanding the role and purpose of data in society (Boyd & Crawford, 2012; Kitchin, 2014). As Van Dijck (2014) argues, dataism, situated between scientific paradigm and ideology, is the current dominant understanding of big data, which itself is '[...] a widespread belief in the objective quantification and potential tracking of all kinds of human behaviour and sociality through online media technologies' (p. 198). Seen as an ideology, dataism supports '[...] a self-evident relationship between data and people, subsequently interpreting aggregated data to predict individual behavior' (Ibid., p. 199). Accordingly, for Van Dijck, dataism is conditioned on the 'trust in the objectivity of quantified methods as well as in the independence and integrity of institutions deploying these methods' (Ibid., p. 204).

Crooks (2017) conceptualises this understanding of the relationship between data and the world as representationalism, i.e. 'the view that representations (here, digital data) stand in a mimetic relation to some external reality from which they are ontologically distinct' (p. 289). However, as his ethnography of data professionals in the field of urban education demonstrates, ensuring the representational capacity of data requires work on behalf of such professionals who deal with the uncertainty and ambiguity of data.

Thus, while a varying degree of uncertainty and ambiguity is present in the data practices of professionals, it is further ideologically masked within dataism as the dominant perspective on data, which positions data as a mimetic representation of the external reality. Yet, while dataism is positioned to be the currently dominant view of data's social role, alternative understandings and practices of working with data coexist alongside it, occasionally challenging dataism's epistemological and political stance as the widespread mode of comprehending the relationship between reality and data. Among these alternative modes of engaging datafication is data activism, which as a concept aims to capture instances of how activists use data for social change and/or to resist aspects of datafication they deem undesirable (Milan & Gutiérrez, 2015).

Data activists' interventions occur on many levels, among which is the cultivation of new 'epistemic cultures', which '[...] point to innovative and potentially transformative ways of relating to big data and its consequences' (Milan & Van der Velden, 2016, p. 63). After elaborating on the meaning of 'alternative epistemologies' of data activism, Milan and Van der Velden (2016) suggest delving deeper into studies of what data means, represents, and questions for activists. Crooks and Currie (2021) also argue that data relations bring both visibility and risks for minoritized communities, suggesting that the meaning of 'data' in activist practices can be further explored for its agonistic potential to advance contestation, rather than resolution of conflicts. From this perspective, data are not a mere unproblematic representation of a pre-given world; rather, as Gray (2018) argues, we may inquire into how data are involved in 'world-making.' Among these ways of world-making with and through data, Gray (2018) suggests thinking of a 'horizon of intelligibility' as a type of data world. By looking at how data are implicated '[...] in making things up and making things intelligible with data,' we may better grasp how 'reality' becomes rendered knowledgeable and actionable for activists who scrutinize, challenge or employ data in their work (Ibid., p. 7).

In social movement research, knowledge has recently also become a scholarly concern (Casas-Cortés et al., 2008; Chesters, 2012; Cox, 2014). Della Porta and Pavan (2017) suggest thinking about the 'repertoires of knowledge practices' of social movements, which are practices used by activists to make sense of their collective self, their networks of action, and political alternatives. Following this line of theoretical thinking, empirical research within social movements studies has examined the interrelation between social movements, digital media, and knowledge (Pavan & Felicetti, 2019; Pavan & Mainardi, 2020). But while it has been argued that activist knowledge predominantly originates in personal experience as 'testimony and storytelling' (Della Porta & Pavan, 2017, p. 11; Esteves, 2008, as cited in) less has been said about how data re-mediates activist knowledge and transforms its alternative epistemologies.

This line of scholarly inquiry within social movement studies may be productively bridged with data activism research. The question of what kind of knowledge about the world does data produce includes understanding that the seemingly unitary and solidified term 'data' in practice encapsulates multiple meanings for activists (Schrock & Shaffer, 2017). Within this line of scholarship researchers have shown how the employment of data in activist work alters the understanding of social change (Gutiérrez & Milan, 2019); how open data initiatives evoke and reconfigure longstanding political ideas and institutions such as democracy (Baack, 2015) or citizenship (Gabrys et al., 2016); or how data are used in a 'counter-data action' manner to express political dissent and challenge the status quo (Currie et al., 2016; Meng & DiSalvo, 2018). This article aims to contribute to studying the intersection between data, knowledge, and activist practices by advancing not the questions of meanings, which activists ascribe to data, but about what kinds of epistemic cultures are cultivated in data activists' practices as a result of the changing relationship between data and reality for activists.

### *Contentious data politics of traffic accidents: the DTP Map as a part of the urban social movement in the Russian context*

Data activists are cultivating alternative epistemologies of data not in the laboratory, but the political field. To account for the contentious politics surrounding data, Beraldo and Milan (2019) build on the work of Tilly and Tarrow (2015) and critical data scholars to propose a theoretical synthesis of social movements studies and data activist research. According to them, the remediation of activism through data creates two closely intertwined in practice, yet analytically separate, modes of data activism. They suggest distinguishing between 'data as stakes', i.e. 'data-oriented activism', and 'data as repertoires' ('data-enabled activism'); moreover, an additional axis of their analysis is the distinction between 'individual' and 'collective' action.

While data-oriented activism considers data infrastructures as a site of contention, data-enabled activism aims to mobilise data to achieve desirable goals of social change; both are better understood as existing on a continuum. As Beraldo and Milan (2019) write, '[. . .] data sometimes intervene as both issues on their own and as a component of a repertoire of action' (p. 8). Going beyond the simple 'means' vs. 'ends' dichotomy, they suggest looking at how within activist initiatives a 'political data subject' comes into being as a result of the interplay between these analytical dimensions. While data-oriented activism can be exemplified by digital rights initiatives and privacy advocates, the instances of data-enabled activism include data journalism that is oriented to advocacy and open data initiatives.

As I shall demonstrate later, the emergent practices of the DTP Map oscillate between comprehending data as 'stakes', on the one hand, and as 'repertoires', on the other. In the process of its constitution, multiple collectives are involved in the mapping of traffic accidents in Russia. Yet while proponents of dataism often speak of changes brought by data in a universalist manner, Milan and Treré (2019) have pointed out the urgent need to move beyond 'data universalism' to consider how the social life of data unravels in contexts different from Western ones, which are often implicitly taken to be the default.

To answer this call, the focus is placed on the Russian context of data activist and civic technology initiatives. It is worth noting that the development of open data policies on the governmental level has been mostly dormant after the shutdown of the Open Government Initiative in 2018. However, during the 2012–2018 period several governmental projects were aimed at promoting governmental openness, freedom of information, and open data repositories.

Despite such efforts, there is evidence of several significant issues regarding how official statistics in Russia are produced. As argued by Molyarenko (2019), the production of official statistics across Russia is accompanied by both unintentional and intentional distortions. While unintentional ones are explained by the overwhelming bureaucratic pressure on ordinary workers in governmental and municipal departments, the intentional misrepresentations of statistical data originate in the fact that the same statistics are used to evaluate the workers' performance. Moreover, following Scott (1998), Molyarenko also emphasises that some phenomena are simply not captured by the statistical bureaus of the Russian state.

These intentional distortions of official data seem to be also present in the case of the official statistics of road traffic accidents. While the more comprehensive analysis of traffic accidents data production remains out of the scope of this article, many chat participants and DTP Map developers themselves have expressed similar concerns. For instance, the official ending of governmental programs, aimed at reducing mortality and the number of traffic accidents, occasionally suspiciously 'coincides' with surprising reductions of bureaucratically unwanted metrics. Traffic accidents mortality is also one of the metrics, according to which the performance of Russian governors is assessed, which in turn creates incentives for distorting the data or making it invisible thereby making the authorities unaccountable.

Moreover, (data) activist initiatives in contemporary Russia exist in the context of widespread scepticism towards mainstream electoral politics. As some scholars and commentators have argued, many people in contemporary Russia are sceptical towards institutionalised politics (Erpyleva & Magun, 2014; also see Weiser & Yudin, 2020). Yet in her study of civic tech projects and hackathons in Russia, Ermoshina (2019) argues that many Russian civic tech activists are less apathetic. In the atmosphere of distrust towards traditional political institutions, grassroots initiatives are '[...] trying instead to invent, experiment, and tinker with new tools – assemblages of programming code and law – that have the potential to fight political apathy and improve the everyday lives of Russian citizens' (Ibid., p. 88). These various civic tech projects can be seen as promoting alternative modes of civic empowerment and political engagement that translate the experience of 'trouble' to the interface of civic applications (see also Ermoshina, 2014, 2019).

## *Epistemological ambiguity of data as a constitutive feature of the DTP Map project*

Data activism has the potential to be a space of inventive alternative epistemologies of data, which are distinct from the dataism and representationalism that accompanies it. However, whether data activists think about knowing the world differently with data, is a question to be empirically examined. In this article, I suggest the term 'the epistemological ambiguity of data' to analyse tensions between data and the knowledge about the world it offers, arguing that it is an always present, collective, and constitutive trait of the DTP Map project. Epistemological ambiguity of data is a concept that originated from studying how the relationship between data, knowledge, and reality is formed in the case of the DTP Map. Yet, despite the concept's grounded nature, in introducing it I hope that it can serve as a heuristic tool for analysing other instances of data activism.

Epistemological ambiguity of data refers to a practical tension between complete trust in data (and, consequently, establishing a mimetic relationship between reality and data) and its total dismissal as a source of knowledge. Such ambiguity originates in the practical challenge of using data for social change while remaining sceptical of how it informs an understanding of reality. While data about social life has often been understood as 'formulas for agreement' (Desrosières, 1998, p. 66), the concept of epistemological ambiguity of data allows us to account for how this agreement is forged and sustained in practice.

Data activists, who partake in making the DTP Map, navigate the complex challenge of striving to gain knowledge about traffic accidents by mapping and examining data about it, while simultaneously staying critical of it. On the one hand, as data are 'never raw' and often even 'dirty', the relation of direct and unproblematic mimetic representation between data and reality cannot be established. On the other hand, to fully dismiss traffic accidents data would mean rendering data activists' efforts in using data to advance social change meaningless. Therefore, epistemological ambiguity of data is a 'middle ground' between data representationalism and the outright dismissal of data as a valuable source of knowledge about reality and the political action it makes possible.

Two characteristics of epistemological ambiguity of data should be distinguished. First, as a practical tension, such ambiguity is only partly and temporarily relieved in data activists' practices, thereby making it a constitutive feature of data activists' projects, rather than simply a temporary limitation to what activists know about data. Since DTP Map continuously updates data on the geoweb platform, it is impossible to claim once and for all that all concerns about new data are gone. Instead, the continual constitution of the project requires that data are always treated in an epistemologically ambiguous way to uphold the project's enduring commitment: activists ensure that data can inform us about reality and in doing so, assist in making social change possible. In these data practices, as I shall demonstrate in the subsequent section of the article, data's epistemological ambiguity is managed through practices which treat it as a stake of contention, thereby allowing it to be employed as a repertoire of contention.

Second, these data activists' practices of dealing with ambiguity are not individual, but rather collective and distributed, enrolling users in the unending constitution of the project. Therefore, to make use of data as a source of knowledge about the world involves securing the trust of users to further advance the cause of diminishing the instances of traffic accidents in Russia. As I shall show in the next section, contentious politics of data, in which data are usually treated either as a stake or a repertoire of social change, are tightly coupled with the epistemological ambiguity of data. When data are treated as a stake, data's mimetic relationship to the external reality needs to be secured by ensuring that it accurately represents things it assumes to stand for. As a result, data can be treated as a repertoire – a mode of advocacy for desired social change. Both project developers and users are enrolled in these data practices and the possibility to do so contributes to sustaining trust in DTP Map.

Overall, epistemological ambiguity epitomises an always present tension between conflicting understandings of data: the conflict between treating data as a mimetic representation of the world and completely dismissing its representational capabilities. Epistemological ambiguity, as it is present in the case of the DTP Map, disturbs the representationalist relation that data are assumed to have with reality under the guise of dataism. Yet, rather than adopting a socio-critical imaginary that is sceptical of possible social change achieved through technology, data activists, who in their practices treat data in an epistemologically ambiguous way, commit to using data for social change without simultaneously adopting a totally techno-optimistic stance towards data (see Lehtiniemi & Ruckenstein, 2018).

The concept of epistemological ambiguity of data makes it possible to address the role of data activism beyond the mere representation of social issues. In the context of data activism research, the non-representational approach to data has been adopted by Rajão and Jarke (2018). By emphasising the productive, rather than the representational capabilities of data, they have studied environmental data activism around deforestation of the Amazon rainforest to argue that it is not sufficient to see data only as a representation of reality since this would unproblematically assume the existence of a pre-given 'outside world', which is to be represented with and through data. Instead, relying on the actor-network theory approach, they suggest considering how particular arrangements of human and non-human actors are involved in multiple enactments of 'Amazon deforestation'. Similarly, instead of thinking about data activism as a mode of mere representation of social issues, the concept of epistemological ambiguity of data provides space for exploring more nuanced relationships between data and the knowledge it offers about the world, which is born out of data activists' epistemic cultures.

In this part of the article, I have conceptualised the case of the DTP Map as a case of alternative data activist epistemic culture, situating its contentious politics in the Russian context. I have introduced the concept of epistemological ambiguity of data to analytically capture the relationship between data as a stake and data as a repertoire as it pertains to the knowledge about traffic accidents that data activists strive to obtain. In the next sections, I introduce the methodological foundations of the article and turn to the closer empirical examination of developers' and users' practices.

## Methodology

In this article, I provide a case study of the DTP Map project by building on the interviews with two key developers (n = 2), as well as a 2-year long digital ethnography of the project's public online chat, its media coverage (n = 10), and other project-related documents (n = 3). These various empirical materials helped to form a holistic understanding of the project as it was discussed in the theoretical part.

The interview guide included questions about the activists' vision of social change that the project brings, engagement with other collectives (e.g. journalists, urban activists, researchers) as well as the activists' data and mapping practices. During the digital ethnographic part of the research, I have attempted to examine user practices as they engage in discussions with each other, discover errors in the data, and suggest improvements for the DTP Map as a project. Media coverage of the projects helped to establish how the DTP Map is perceived by the public while project-related documents, such as GitHub documentation and internal research conducted by the developers, helped to better grasp how the project transformed with time.

As discussed before, some scholars within social movement studies have started to consider activists' alternative epistemologies and see them as knowledge producers, rather than objects of scholarly inquiry (Chesters, 2012; Kazansky et al., 2019). Against the background of understanding social movements as knowledge-producers, in my research I have attempted to simultaneously pay attention to how data are both leveraged to produce knowledge about political alternatives and restrain from considering this case as a 'data-driven' instance of activism, i.e. primarily focusing on what is the role of 'data' in geoweb mapping practices (Della Porta & Pavan, 2017). Instead, I have methodologically attempted to decenter datafied technologies and consider how within the DTP Map project the relationship between knowledge, reality, and data are established and sustained. In the case of exploring alternative epistemologies of the DTP Map project, considering map-making as part of activist practice also entails seeing them as a form of knowledge creation (Kennedy et al., 2016).

## The case of the DTP Map: the users' trust and the interplay between data as stakes and as a repertoire

The DTP map project was started by Sergey Ustinov and Alexey Radchenko in 2018; Ustinov was the lead developer in the project's development while Alexey was an urban activist. Initially, the interest in the project came from Radchenko's engagement with the issues of safety of the urban environment in Moscow. As he narrates, the Moscow Department of Transport's official position on urban safety was that overpasses and underpasses make the city safer in terms of traffic accidents, while Alexey's understanding, expertise, and experience led him to believe that the opposite was happening.[1] To provide a more definitive argument on this issue, he needed to acquire 'decisive data-based evidence' of his claims that he could demonstrate publicly.

The DTP Map project eventually extended beyond Moscow to cover other Russian cities and regions with a geoweb map of traffic accidents, based on the official GIBDD statistics. The project is executed in line with the concept of Vision Zero – a multinational project which aims at reducing and eliminating traffic accidents. Originally started in Sweden, Vision Zero's ideas reached many countries around the globe. Moreover, from the very beginning, Vision Zero has relied heavily on data-based evidence of traffic accidents, without which the argument for proper intervention cannot be made.

At the same time, what has complicated the understanding of traffic accident data as a purely unquestionable representational resource and repertoire for social change was its partial unreliability. The unreliability of official data on traffic accidents manifests itself in a twofold manner. First, it is conditioned on the unintentional and intentional distortions of statistics production, as outlined by Molyarenko (2019) and presented above.

Second, the data are occasionally incomplete or obviously 'wrong', as was revealed in data and mapping practices of developers and users. For instance, the geolocation of traffic accidents sometimes is misplaced (most obviously when a traffic accident happened in the river, according to data) or when notable cases of traffic accidents, which are covered by media, are not included in the official datasets, to mention a few such instances. Consequently, data activists cannot take government data for granted; it must be interrogated and questioned to be further used as a repertoire of social change.

Taking these features of official data into account also has epistemological consequences for the DTP Map as a project. How do you provide a map of traffic accidents for urban activists, journalists, and analysts to further act on, if you also have to consider that the data itself may be partially unreliable?

In the previous section, I have conceptualised this practical dilemma as 'epistemological ambiguity of data', which simultaneously refers to the activists' desire to enhance and re-affirm the data's representational capabilities, while also attempting to account for the 'ruptures' between reality and data. Far from both dataism – which in its pure form seems to exclusively promote data's representational capacities without accounting for the conditions of its production, distribution, and usage – and the outright dismissal of data on grounds of its dubious character, this epistemological ambiguity of data is accounted for in the DTP Map project in the mapping and data practices, in which multiple collectives engage.

I argue that the epistemological ambiguity of data both reflects a practical tension for the project, and, at the same time, constitutes the project as a whole. It can be only temporarily mitigated and remains always present as the project proceeds.

To analytically separate two sets of data practices, I now continue to describe the practices in which activists and collectives engage to consider data as a stake in their work. Then I consider how such practices contribute to advancing the role of data as a repertoire of social change.

### *Data as a stake of grassroots traffic accidents mapping: accountability from below and practices of curating data*

Data becomes a stake in activist practices in three ways. First, developers and various people, involved in the project on an occasional basis, such as urban activists and other concerned citizens, write official appeals to GIBDD in which they describe the data inaccuracies. For instance, they write to GIBDD about missing traffic accidents that they know of or misplaced geolocations. According to the DTP Map developers, the departmental staff within GIBDD, which is responsible for publishing traffic accident data, was initially surprised that someone was using this data. Yet as time passed, the interaction between this department and the concerned public changed, as the former started to amend data errors more eagerly.

The developers assisted in the process of writing such complaints by providing users with a more standardised template that was shared in their online chat. According to data activists, the 'quality of data' (as they consider it in their work) eventually improved for reasons unknown to them, though no structural changes to the production of official statistics seem to have happened. These collectively distributed data practices can be seen as accountability from below, in which the concerned public is engaging data as a stake by closely inspecting its relation to the represented reality.

Second, the DTP Map developers themselves, with the help of other volunteer programmers, fix data errors themselves by adjusting coordinates or fixing outright wrong data values (such as one man having 95 years of driving experience). According to developers, fixing such errors is crucial for maintaining the users' trust in their project; I shall come back to the idea of trust between the project and users later. Third, developers write public posts in Telegram channels and blog posts, with suggestions on how to analyse data by paying close attention to the limitations of data; they generally highlight that the data cannot be taken for granted or elaborate on what cannot be inferred from data. As Radchenko comments on this issue in an interview:

> When DTP Map appeared, activists, journalists, and the authorities – all of them began to rely on these data points as something, I don't know, 'one hundred percent truthful'. As I have said before, it is difficult to argue with numbers, and for an official, it is difficult to counter numbers; they are often an ironclad argument in your favour. But numbers can be manipulated, and when I saw how much trust these numbers have gained, it became clear that more complex rules should be introduced to make clear that data can be interpreted differently, and that data can be wrong.

In turn, Ustinov capaciously remarked that data should be understood 'with reservations'. For developers, such reservations entail paying closer attention to what data stands for, the possibility of data errors or inaccuracies, and accounting for broader political accountability relations of which different municipal and governmental authorities are a part. Thus, data practices require curation, guidance, and contextualisation, which in turn can be comprehended as accounting for aspects of a social world that shape data production.

### *Data as a repertoire: re-defining responsibility and engaging regional communities*

I now turn to consider how data are seen as a repertoire of social change in the case of the DTP Map project. I discuss this aspect of the project in a twofold manner. First, I illuminate how the changes to the crowdfunding model of the project have emphasised the regional dimension of traffic accidents as opposed to GIBDD's administrative district scale. Additionally, I examine how the DTP Map was envisioned by the developers and further used by urban activists, journalists, and researchers as an advocacy tool aimed at reshaping the urban environment. Second, I address how the DTP Map shifts the understanding of responsibility for traffic accidents from the traffic participants themselves to the urban environment and road infrastructure.

Initially, the DTP Map emerged with only data on traffic accidents in Moscow, eventually covering more and more Russian regions and cities. The developers of the DTP Map intended that local urban activists, journalists, and local communities will use their map to address issues of traffic accidents most relevant to them as users. Rather than thinking of producing a map to just be put out there, the developers were reflexive about how it could be put into action by others. Thus, the mapping practices reflected the project's vision of social change. The DTP Map turned to the crowdfunding model by asking people to donate a small sum of money for a particular region, which became successful as compared to the previous model of donating to the project as a whole. After the money had been gathered, the developers added regional and city data to the map. Once this had happened, they contacted local journalists to ask for press coverage of a region's presence on the DTP Map with occasional success.

This crowdfunding mechanism helps to produce traffic accidents as an issue on a regional scale by linking together data as a repertoire of social change and issues of particular urban communities. For instance, activists in the urban social movement Krasivy Kirov (Красивый Киров; Beautiful Kirov), who have long been preoccupied with issues of the urban environment in Kirov, have used the map and data on traffic accidents as an argument in their discussions with the city administration on the quality of the roads in the city. As the coordinator of Krasivy Kirov, Igor Bagin commented in the local media '7x7': 'I do not know if this has anything to do with our publications on this subject, but now officials have begun to install speed bumps near crosswalks and install pedestrian fences during road repairs'. The developers have also written an article on the well-known Russian website for the IT community – habr.com – in which they described how to conduct research on traffic accidents in a certain region using their map and further advising to send this research to local news outlets to potentially influence responsible administrators (Ustinov, 2019). Similar cases of the DTP Map used as an argument in activists' negotiations with the city administration were documented by developers in at least Chelyabinsk and Yuzhno-Sakhalinsk. Some individuals in various Russian cities have also used the Map to write complaints to the authorities.

Moreover, the translocal, distributed, and participatory character of the DTP Map predisposed people to write in the project's Telegram chat about the notable missing data of media-covered traffic accidents, to report misplaced geolocation, or to point out other data or map inaccuracies. This could not be achieved single-handedly by the project developers since it required local knowledge of different Russian cities' streets, daily embeddedness in the local media environment, and additional time.

Additionally, coming from an urban activism standpoint, the DTP Map developers emphasised the role of infrastructure and the urban environment in their public statements on the causes behind traffic accidents. Following the concept of 'Vision Zero', they posit that traffic participants 'should be allowed to make mistakes', and that road infrastructure should be configured in a way that allows for such mistakes. Thus, the project's end goal is not to limit the number of fatal road traffic accidents but even potentially eliminate them through changes to the urban environment and road infrastructure.

The DTP Map developers also work closely with City Projects (Городские проекты; Gorodskie proekty), which is a non-profit organisation with branches in various Russian cities, founded by well-known Moscow urban activists Maksim Katz and Ilya Varlamov. City Projects aims to make the urban environment safer and more convenient for its inhabitants. This connection between the DTP Map and City Projects allows us to see the former as part of a broader urban social movement in contemporary Russia, envisioned to transform city landscapes (Zhelnina & Tykanova, 2019).

The opposite of such a vision is the officially declared position of the Ministry of Internal Affairs, of which GIBDD is a part. In the official strategic document 'Road Safety Strategy in the Russian Federation for 2018 – 2024' one can read the following statement: 'Thus, the state of road traffic safety is largely determined by the discipline and level of professional skill of drivers' (Rossiyskaya Gazeta, 2018). One of the federal-level projects, launched within the framework of the Strategy, is called 'Culture on the Roads', which aims to address audiences of different socio-demographic groups to promote '[...] awareness of each person's role and responsibility for what happens on the road' (Ministry of Internal Affairs, 2018). Thus, on some level the official position on traffic accidents seems to place responsibility at a personal level, emphasising the accountability of an individual. Contrary to such understanding, many urban activists in Russia, including the DTP Map developers, redefine responsibility behind traffic accidents, turn the focus onto the infrastructure and urban environment, and suggest thinking about them as the legitimate points for political and governmental intervention, instead of attempting to cultivate a specific type of individual road awareness.

As I have demonstrated, the DTP Map project combines an understanding of data as both a stake and a repertoire of social change. Moreover, these are not in practice separated treatments of data by activists, but rather they are deeply interrelated for developers through a key category of users' trust in the project. According to developers, users must trust the data, on which the project relies, to trust the project as a whole. In other words, if data errors, which are obvious to the users, are discovered, trust in the project can no longer be maintained. Radchenko comments on the importance of users' trust in the following way:

> It is not advantageous for us if these data are no good, because then the level of trust in us, in Vision Zero, in the whole urban movement also decreases. On the one hand, we are not responsible for the data because GIBDD is the one producing and accountable for it, but on the other hand, the responsibility is on us . . . We are a 'showcase' of this data, we speak on the basis of it.

On the one hand, trust in the project is grounded in the fact that the data on traffic accidents are predominantly collected by the government – also an important feature of data, when considering negotiations with the city administration. On the other hand, the curating work of the developers also contributes to sustaining users' trust because activists also review for possible errors in data such as misplaced geolocations or missed traffic accidents. As I have described earlier, users also treat data as a stake making this process of accounting for data collective, continual, and distributed.

## Conclusion

In this article, I have explored the case of the DTP Map – the interactive geoweb map that uses government data on traffic accidents to make it more accessible to the public and address issues of road infrastructure and the urban environment as central causes of traffic accidents in Russia. Employing the theoretical framework of contentious data politics, I have tried to understand how activists comprehend the epistemological status of data. The latter shifts as data become a repertoire or a stake in the activists' project. At times, data are employed to demonstrate traffic accidents as an urban infrastructural issue and make arguments for changes (as it is in urban activists' negotiations with the city administrations), at other times, it is doubted, questioned, corrected (as it is in addresses to GIBDD about data inaccuracies.)

This article contributes to the interdisciplinary studies of social movements and data activism by conceptually and empirically articulating the relationship between knowledge, activism, and data practices, enhancing scholarly understanding of activist alternative epistemologies. Additionally, this case also enriches our understanding of data and digital activism in the Russian context (Lonkila et al., 2021). The previous research on Russia has explored how activists employ digital technologies for social change (Ermoshina, 2014, 2019) and how they manage their visibility within the online environments considering possible opportunities and risks (Lokot, 2018). Navigating political and institutional obstacles, civil society actors not only contest state narratives on personal data (Lokot, 2020) but also, as I argue in this article, actively cultivate alternative data epistemologies.

Thus, what makes data activist epistemology 'alternative' to dataism, in this case, seems to be that the official data on traffic accidents do not possess epistemological coherency outside of users' or developers' practices. From the developers' perspective, the users' trust must be sustained, and so data must always be questioned and interrogated; at the same time, as the project emerged within the broader agenda of Russian urban activism, the employment of data as a repertoire of social change has been crucial to the DTP Map's goals from the beginning. Rather than resolving the epistemological ambiguity of data altogether to make sense of it as 'objective' or 'fully representative' of traffic accidents in Russia, this ambiguity is constantly reproduced in data and mapping practices, as the DTP Map project oscillates between enacting data as a repertoire in one set of practices and as a stake in others.

Such a conclusion begs further questions about the relationship between data, knowledge, and activism. First, in considering social movements and activists as knowledge producers, rather than purely objects of research, can we further inquire into what kind of epistemic cultures they cultivate and what sets of practices are enrolled in these processes, both across time and space? Second, what kind of activists' relationship to knowledge emerges when 'data' consists of several resources (e.g. quantified data, testimonies), rather than from a single one? Third, if retaining epistemological ambiguity of data in data practices is fundamental to trust between users and developers in the case of the DTP Map, as I argue, what kinds of repertoires of knowledge practices are instrumental for supporting and sustaining social movements and activists' projects or, vice versa, detrimental for activists and their movements?

## Note

1. Ustinov and Radchenko have been interviewed for this research and wished for their names to be public in the article. After discussing it with them, I have tried to carefully consider the potential benefits and risks of such publicity and in the end, reasoned it to be the best possible decision for the informants.

## Acknowledgements

The author thanks three anonymous reviewers for their insightful feedback and special issue organisers for making the issue possible. Additionally, the author is grateful to Polina Kolozaridi and Leonid Yuldashev who provided generous feedback during various stages of the project.

## Disclosure statement

No potential conflict of interest was reported by the author(s).

## ORCID

Dmitry Muravyov http://orcid.org/0000-0002-8568-2199

## References


Baack, S. (2015). Datafication and empowerment: How the open data movement re-articulates notions of democracy, participation, and journalism. *Big Data & Society*, *2*(2), 1–11. https://doi.org/10.1177/2053951715594634

Beraldo, D., & Milan, S. (2019). From data politics to the contentious politics of data. *Big Data & Society*, *6*(2), 1–11. https://doi.org/10.1177/2053951719885967

Boyd, D., & Crawford, K. (2012). Critical questions for big data: Provocations for a cultural, technological, and scholarly phenomenon. *Information, Communication & Society*, *15*(5), 662–679.

Casas-Cortés, M. I., Osterweil, M., & Powell, D. E. (2008). Blurring boundaries: Recognizing knowledge-practices in the study of social movements. *Anthropological Quarterly*, *81*(1), 17–58.

Chesters, G. (2012). Social movements and the ethics of knowledge production. *Social Movement Studies*, *11*(2), 145–160.

Cox, L. (2014). Movements making knowledge: A new wave of inspiration for sociology? *Sociology*, *48*(5), 954–971.

Crooks, R. (2017). Representationalism at work: Dashboards and data analytics in urban education. *Educational Media International*, *54*(4), 289–303.

Crooks, R., & Currie, M. (2021). Numbers will not save us: Agonistic data practices. *The Information Society*, *37*(4), 1–19.

Currie, M., Paris, B. S., Pasquetto, I., & Pierre, J. (2016). The conundrum of police officer-involved homicides: Counter-data in Los Angeles County. *Big Data & Society*, *3*(2), 1–14. https://doi.org/10.1177/2053951716663566

Della Porta, D., & Pavan, E. (2017). Repertoires of knowledge practices: Social movements in times of crisis. *Qualitative Research in Organizations and Management: An International Journal*, *12*(4), 297–314.

Desrosières, A. (1998). *The politics of large numbers: A history of statistical reasoning*. Harvard University Press.

Ermoshina, K. (2014). Democracy as pothole repair: Civic applications and cyber-empowerment in Russia. *Cyberpsychology: Journal of Psychosocial Research on Cyberspace*, *8*(3), 24–37.

Ermoshina, K. (2019). For code and country: Civic hackers in contemporary Russia. In M. Biagioli & V. A. Lépinay (Eds.), *From Russia with code: Programming migrations in post-Soviet times* (pp. 87–109). Duke University Press.

Erpyleva, S., & Magun, A. (Eds.). (2014). *Politika Apolitichnyh: Grazhdanskiye Dvijeniya v Rossii 2011–2013 godov [Politics of the apolitical: Civic movements in Russia in 2011 2013]*. Novoye Literaturnoye Obozreniye.

Esteves, A. M. (2008). Processes of knowledge production in social movements as multi-level power dynamics. *Sociology Compass*, *2*(6), 1934–1953.

Gabrys, J., Pritchard, H., & Barratt, B. (2016). Just good enough data: Figuring data citizenships through air pollution sensing and data stories. *Big Data & Society*, *3*(2), 1–14. https://doi.org/10.1177/2053951716679677

Gray, J. (2018). Three aspects of data worlds. *Krisis: Journal for Contemporary Philosophy*, (1), 3–17.

Gutiérrez, M., & Milan, S. (2019). Playing with data and its consequences. *First Monday*, *24*(1). https://doi.org/10.5210/fm.v24i1.9554

Hacking, I. (1990). *The taming of chance*. Cambridge University Press.

Kazansky, B., Torres, G., van der Velden, L., Wissenbach, K., & Milan, S. (2019). Data for the social good: Towards a data-activist research agenda. In D. A. K. Devitt & M. Mann (Eds.), *Good data* (pp. 244–259). Institute of Network Cultures.

Kennedy, H., Hill, L., Aiello, G., & Allen, W. (2016). The work that visualisation conventions do. *Information, Communication & Society*, *19*(6), 715–735.

Kitchin, R. (2014). Big Data, new epistemologies and paradigm shifts. *Big Data & Society*, *1*(1), 1–12. https://doi.org/10.1177/2053951714528481

Lehtiniemi, T., & Ruckenstein, M. (2018). The social imaginaries of data activism. *Big Data & Society*, *6*(1), 1–12. https://doi.org/10.1177/2053951718821146

Lokot, T. (2018). Be safe or be seen? How Russian activists negotiate visibility and security in online resistance practices. *Surveillance & Society*, *16*(3), 332–346.

Lokot, T. (2020). Data subjects vs. people's data: Competing discourses of privacy and power in modern Russia. *Media and Communication*, *8*(2), 314–322.

Lonkila, M., Shpakovskaya, L., & Torchinsky, P. (2021). Digital activism in Russia: The evolution and forms of online participation in an authoritarian state. In Gritsenko, D., Wijermars, M., Kopotev, M. (Eds.),*The Palgrave handbook of digital Russia studies* (pp. 135–153). Palgrave Macmillan.

Meng, A., & DiSalvo, C. (2018). Grassroots resource mobilization through counter-data action. *Big Data & Society*, *5*(2), 1–12. https://doi.org/10.1177/2053951718796862

Milan, S., & Gutiérrez, M. (2015). Citizens' media meets big data: The emergence of data activism. *Mediaciones*, *11*(14), 120–133.

Milan, S., & Treré, E. (2019). Big data from the South (s): Beyond data universalism. *Television & New Media*, *20*(4), 319–335.

Milan, S., & Van der Velden, L. (2016). The alternative epistemologies of data activism. *Digital Culture & Society*, *2*(2), 57–74.

Ministry of Internal Affairs (2018, November 12). Мурманская область присоединяется к федеральной кампании «культура на дорогах» [The Murmansk region joins the federal program "Culture on the Road"]. https://мвд.рф/news/item/14947221

Molyarenko, O. (2019). Formirovanie gosydarstvennoy statistiki: Vzglyad snizy [Generating state statistics: a view from «below». *ЭКО [EKO]*, *10*(544), 8–34.

Pavan, E., & Felicetti, A. (2019). Digital media and knowledge production within social movements: Insights from the transition movement in Italy. *Social Media+ Society*, *5*(4), 1–12.

Pavan, E., & Mainardi, A. (2020). At the roots of media cultures. Social movements producing knowledge about media as discriminatory workspaces. *Information, Communication & Society*, *23*(12), 1738–1754.

Porter, T. M. (1986). *The rise of statistical thinking, 1820-1900*. Princeton University Press.

Rajão, R., & Jarke, J. (2018). The materiality of data transparency and the (re) configuration of environmental activism in the Brazilian Amazon. *Social Movement Studies*, *17*(3), 318–332.

Rossiyskaya Gazeta (2018) Стратегия безопасности дорожного движения в российской федерации на 2018 - 2024 годы [Road safety strategy in the Russian federation for 2018-2024]. https://rg.ru/2018/01/24/strategiya-site-dok.html

Schrock, A., & Shaffer, G. (2017). Data ideologies of an interested public: A study of grassroots open government data intermediaries. *Big Data & Society*, *4*(1), 1–10. https://doi.org/10.1177/2053951717690750

Scott, J. C. (1998). *Seeing like a state: How certain schemes to improve the human condition have failed.* Yale University Press.

Tilly, C., & Tarrow, S. (2015). *Contentious politics* (2nd ed.). Oxford University Press.

Ustinov, S. (2019), February 3 Карта ДТП [DTP Map]. Habr. https://habr.com/ru/post/438700/

Van Dijck, J. (2014). Datafication, dataism and dataveillance: Big data between scientific paradigm and ideology. *Surveillance & Society*, *12*(2), 197–208.

Weiser, T., & Yudin, G., Eds. (2020). The public sphere in Russia between authoritarianism and liberation [special issue]. *Javnost - the Public*, *27*(1), 1–96.

Zhelnina, A., & Tykanova, E. (2019). Formal'nyye i neformal'nyye grazhdanskiye infrastruktury: Sovremennyye issledovaniya gorodskogo lokal'nogo aktivizma v Rossii [Formal and informal civic infrastructure: Contemporary studies of urban local activism in Russia]. *Zhurnal Sotsiologii I Sotsialnoy*, *22*(1), 162–192.

# Coordinating and doxing data: Hong Kong protesters' and government supporters' data strategies in the age of datafication

Yao-Tai Li and Katherine Whitworth

**ABSTRACT**

Scholars across various disciplines have discussed the role of ICTs and social media in contentious politics and social movements. Other studies have also highlighted the duality of data particularly in the context of contentious politics: data is simultaneously an object of political struggle and a weapon in repertoires of political struggle. Taking the protesters' collective and networked program of sousveillance during the 2019 Hong Kong Anti-ELAB Movement as an example, this article examines protesters' and pro-establishment's engagement with data, the datafication of police as proxies of the state, and attitudes towards various forms of 'veillance'. The doxing at the heart of this case study provides an example in which data as constituents of movement repertoire and data as a contentious issue in its own right are conflated. This article bridges the surveillance and social movement literature by examining data practice and doxing during the protest.

## Introduction

Hong Kong offers a unique case to consider the intersection of data and contentious politics. The city's postcolonial history has positioned the jurisdiction as an 'open' exception with reference to a 'closed' China. However, in contrast to the narrative of an 'open, smart city', the legislative framework inherited from the British colonial government defining the role of data and guaranteeing rights to information remains weak (Hoyng, 2021; Meng et al., 2019). Access to data has been framed as a service, rather than a right by the Hong Kong government, and as a privilege by private sector corporations (Hoyng, 2021). The resulting information asymmetry has led to Hong-Kong-based data intermediaries to engage in 'counter-data action' by building their own datasets (Meng et al., 2019). In the context of increased tension over Hong Kong's postcolonial political and cultural identity, cross-border data mobility under China's Greater Bay Area plan[1] and the vague definition of 'personal data'[2], activists and citizens have begun to exploit, contest and reimagine the norms around the openness of data.

Most recently, this contestation over data in Hong Kong took place during the Anti-Extradition Law Amendment Bill (Anti-ELAB) movement of 2019.[3]

The networked and datafied contentious performances seen during the movement provide a unique context to re-examine the nexus between data-driven governance and individual data practice during times of contention. The Hong Kong government adopted a hybrid strategy of repression and policing that blended 'hard' and 'soft' methods (Wang et al., 2020). 'Hard strategies' included the forceful suppression of rallies with tear gas and water cannon. 'Soft strategies', on the other hand, included the collection of protesters' data through facial recognition software (Mozur, 2019), transport usage records and social media posts. In response to datafication and surveillance carried out by the state (and pro-government supporters), protesters engaged in anti-surveillance practices such as wearing masks, destroying CCTV cameras, and disabling the facial recognition login on their phone to disrupt the government's ability to collect biometric information. They also appropriated the techniques of the state through an organized program of doxing – the public exposure of private information without consent – that crowdsourced and publicized information about police and people suspected of being 'snitches'. This weaponization of data influenced the movement's dynamic and the contentious performances exchanged between the state and protesters. Police started wearing masks and stopped prominently displaying their badge numbers to conceal their identity from protesters (Mozur, 2019). Government supporters also collected and publicized images of individuals participating in illegal activities such as 'unsanctioned' protests and the destruction of property on platforms such as hkleaks.pk and reboot-thug-archive. The doxing campaigns seen in the Hong Kong Anti-ELAB protests of 2019 serve as a case study that neatly encapsulates the duality of data – that data is simultaneously an object of political struggle and a weapon in repertoires of political struggle (Beraldo & Milan, 2019; van der Graaf et al., 2016).

This article sets out from the assumption that datafication now also exists as a foundational structure within society that can create and remove opportunities for change, particularly during episodes of contention. We situate doxing as a practice that sits at the nexus of Beraldo and Milan's (2019) data as weapon/data as stakes dichotomy. This dichotomy is particularly relevant in the Anti-ELAB movement case because we see an overlap in both means and objects of datafied contention between the two politically opposed groups. We explore the specific dynamics of doxing as an individual and a collective performance within Hong Kong's contentious repertoire. Doxing, when done collectively in the context of contentious politics, is shown to shift power dynamics between claimants, as doxers take on the practices of (or even the role) of the state. Thus, in certain contexts, doxing can become an organized program of either 'sousveillance' (the surveilled subject is 'gazing back at (state) power from below') or 'surveillance' depending on the ideological positioning of the doxers vis-a-vis the state. This assertion frames doxing, typically seen as an unethical practice, as something that can be justified, even morally appropriate, in certain circumstances. Making a judgement on this ethical paradox, as well as qualitatively interrogating the way Hong Kongers perceive this paradox, are beyond the scope of this study and warrant further research.

We focus on the discourses about doxing police and their families on LIHKG[4], as well as two Telegram channels (Dadfindboy and Youcangotojail) organized by protest sympathizers and government supporters respectively to examine the various strategies, target audiences, and moral justifications given for doxing. Linkages between these moral justifications and the persuasive functions of movements (Benford & Snow, 2000; Stewart et al., 2007) can be drawn – which further connects individual instances of doxing as a data practice with collective identity in the movement (Milan, 2015; Trottier, 2017). Our findings suggest that doxing in the Hong Kong example changed the power relations between police, protesters, and government supporters; it also became forms of sousveillance and surveillance that blended more traditional image-based monitoring techniques with contemporary techniques of datafication. We conclude by discussing the consequences of connecting individual data practices with the proactive mobilization of collective identity and the (re)definition of moral boundaries.

## Literature review

### *Contentious data politics*

Studies of contentious politics and data activism have shed light on the importance of agency, power, and practice in the digital age (Beraldo & Milan, 2019; Flesher Fominaya & Gillan, 2017; Pellegrino et al., 2019). Contentious politics refers to when 'ordinary people – often in alliance with more influential citizens and with changes in public mood – join forces in confrontation with elites, authorities, and opponents' (Tarrow, 2011: 7). It can occur over any issue and manifest itself through a variety of mediums, including data. Data are contentious in nature (Beraldo & Milan, 2019; Milan & van der Velden, 2016) and are often repurposed for various political purposes such as mobilization, surveillance and repression (Dencik et al., 2016). Beraldo and Milan's (2019) claim a paradigm shift has occurred in datafication – from data politics to the contentious politics of data. They (2019: 2) define the latter as:

> The multiplicity of bottom-up, transformative initiatives interfering with and/or hijacking dominant, top-down processes of datafication, by contesting existing power relations and narratives and/or by re-appropriating data practices and infrastructure for purposes distinct from the intended.

In this sense, data activism can serve as a 'heuristic tool' that can be used to explore how activism evolves in response to big data (Milan & van der Velden, 2016, p. 66). The term 'data activism' has emerged to describe the adoption of data to reach movement goals or to engage in contestation (Kazansky et al., 2019; Lehtiniemi & Haapoja, 2020). Data activism can take many forms and be 'pro-active' as well as 're-active'. The former refers to instances where data are used as part of civic engagement, advocacy, and campaigning for social change, whereas the latter focuses on how individuals and groups resist corporate intrusion into their private lives, datafication and state surveillance (Pellegrino et al., 2019; Schrock, 2016). Doxing can be both a pro-active and re-active form of data activism and has increasingly been seen in contentious events of recent years such as the Arab Spring, the Black Lives Matter movement and the Anti-ELAB movement of Hong Kong. In these contexts, doxing is a practice grounded in datafication and thus inextricably linked to surveillance.

### *Surveillance, anti-surveillance and sousveillance*

To connect surveillance studies and social movement research, we need to first start with studies on policing protest (della Porta et al., 2006; Earl & Soule, 2006; Jackson et al., 2019). Surveillance – or 'watching from above'—is typically understood as a form of disciplinary power wielded by the state or capital (Foucault, 1977). With specific reference to surveillance and social movements, research has focused on the repression-protest nexus (Jackson et al., 2019; Schlembach, 2018), in other words how policing and protester tactics mutually shape each other and political opportunity structures. Studies have explored the effects, real and perceived, that surveillance has on human behavior and mobilization. Many (e.g. Ullrich & Wollinger, 2011) have highlighted the connection between prominently displayed surveillance devices such as CCTV cameras and compliance with the law. Others (Lyon, 2018; Stoycheff et al., 2019) have considered how less visible forms of surveillance, such as network infiltration and the datafication of individuals (whether it be by states or big-tech), have prompted self-censorship and thus had a 'chilling effect' on the mobilization of dissent. Individuals are shown to be enmeshed within a 'surveillance assemblage' (Haggerty & Ericson, 2000; Ullrich & Knopp, 2018), comprising multiple layers of surveillance actors, including the government, police, their peers, opponents and the general public. Peer-to-peer surveillance has become increasingly easy with advancements in personal technological devices and is particularly important in the context of contention over national (or in Hong Kong's case, local) identity, as it connects popular (people-driven) surveillance practices to state power (Bitso et al., 2012).

Efforts to resist surveillance and equalize embedded power asymmetries have also become popular topic of research. Some (Dencik et al., 2016; Hermida & Hernández-Santaolalla, 2018) have focused on *anti-surveillance*, whereby individuals adopt practices that disrupt or evade the government's attempts to collect their information. Scholars (Maragkou, 2019; Wood & Thompson, 2018) have discussed the role of crowdsourced information which 'facilitate[s] surveillance discovery, avoidance, and counter-surveillance.' Others (Mann & Ferenbok, 2013; Mann, 2005, 2013) in the field of surveillance studies have highlighted the emergence of a new kind of 'veillance'—*sousveillance*—whereby the subject is 'gazing back at (state) power from below' in order to document misconduct and foster accountability. To date, literature concerned with sousveillance has focused on wearable devices that capture, process, store, recall and transmit human sensory information and thus has predominantly considered the individualized nature of sousveillance as a contentious performance. This article provides an example of sousveillance that is focused on a practice not considered as such before and highlights how it can be a highly coordinated, networked program that transcends individuals, and further constitutes a movement repertoire in influencing collective identity. Furthermore, we extend discussion of how sousveillance (and surveillance) has been integrated into mobilization framing and persuasion (Benford & Snow, 2000; Stewart et al., 2007).

### *Doxing and data ethics*

Doxing refers to the public exposure of private documents and information without consent (such as photos, phone numbers, and home addresses) online, with the intent to humiliate, harass, intimidate, punish and/or blackmail targets, or condemn certain actions and ideas (Carty & Reynoso Barron, 2019; Fish & Follis, 2016). In light of the damaging effects doxing can have on its targets, it is typically seen as unethical (McNealy, 2018). However, doxing has been increasingly connected with digital activism and framed as a technique sometimes necessary to reveal wrong-doing or hold wrong-doers to account (Douglas, 2016). The practice enables users to redefine what counts as acceptable conduct and contest the boundaries of what the state defines as acceptable intervention. From this perspective, Trottier (2017) describes certain kinds of doxing as 'digital vigilantism', a user-led policing that defies, distrusts, or supplements police and the state who are unable or unwilling to prosecute perceived offenses. In this framework, the goals of doxing are broadened to include responding to crimes, preventing offenses, providing information, and negotiating values.

To navigate between 'justifiable' and 'unjustifiable' instances of doxing, scholars have developed several typologies that categorize the intention behind doxing, the type of information leaked and the consequences the information release will have on the doxed subject. Douglas (2016), for example, distinguished between doxing that de-anonymizes, targets, and de-legitimizes its subject. In this framework, de-anonymizing doxing publicizes personal information that establishes the identity of a previously anonymous person; targeting doxing refers to releases of information about an individual's circumstances that are not publicly shared; and lastly, de-legitimizing doxing discloses intimate information that harms the integrity of an individual. Trottier (2017) categorizes doxing along several axes including: doxer, target, and/or law enforcement. MacAllister (2017) conceptualizes the purpose of doxing from a legal standpoint. She suggests three functions: to 'punch down' a target either as a symbolic act, to enact revenge or punishment, or simply for the 'fun' of it; to increase the transparency of an issue or group to the public, heightening attention to newsworthy information; or to expose a perceived injustice.

## Data and methods

Following Douglas (2016) and Trottier's (2017) definitions that characterize doxing as the release of private information on a public platform, we chose two platforms – LIHKG and Telegram – commonly used by movement and government sympathizers to 'name and shame' perceived opponents. LIHKG was launched in Hong Kong in November, 2016. It is a Reddit-like, multi-category discussion forum that allows people registered with an institutional or ISP email (not gmail or hotmail) to sign up and anonymously post, upload, vote, and chat. On LIHKG, users can vote discussion threads up or down and thereby popularity drives which threads will be presented first. During the Anti-ELAB Movement, LIHKG became popular amongst protesters to exchange ideas, disseminate information, and make decisions (Lee, 2020). Telegram, by contrast, is a not-for-profit, encrypted, cloud-based messaging app. It distinguishes itself from other platforms such as Facebook and Google by emphasizing its commitment to privacy, and is therefore popular among protesters to share information and organize activities (Maragkou, 2019).

We examined posts on LIHKG specifically related to doxing to get a sense of discourse on doxing the police and their families. To gain insight into how protesters and government supporters actually doxed their perceived opponents, we chose to further investigate posts on two Telegram channels created with the stated purpose to dox perceived opponents: Dadfindboy[5] and Youcangotojail channel[6]. Dadfindboy was established on 7 July 2019, with the stated purpose of publicly releasing the private information of police officers to de-anonymize them. Youcangotojail channel was established on 24 November 2019. The stated goal of Youcangotojail channel is to take collective action against 'rioters' because their practices are 'illegal' and 'disturbing the social stability'.

Protest-related posts on LIHKG and Telegram are predominantly written in colloquial Cantonese, the language spoken in Hong Kong. This lowers the barrier of entry for Hong Kongers to participate in conversations on the protests while simultaneously limiting those who cannot read traditional Chinese characters from engagement. This infrastructure provides what Tufekci (2017) describes as 'network internalities', the benefits and collective capabilities obtained from the constant work of negotiation and interaction required to maintain the networks as functioning and durable social and political structures.

Methodologically, first, we registered an account with each platform to review the content available and understand any differences between publicly available information and the information available to members only. We concluded that the content available to members and non-members is similar in the channels this research focuses on. Based on the linguistic conventions adopted by users (e.g. 'dog' is a common usage referring to the police), we scraped LIHKG posts that included the following keywords in their titles: 'dadfindboy', 'bad cop + doxing', 'cop's wife', 'cop's son/daughter', 'bad cop + punish', 'fifty cents army'[7], 'dog's badge number', 'dog's home', 'dog's name', among others. A total of 154,920 posts were scraped. We then used a data analytic program, KNIME[8], to filter out duplicated posts, posts that only contained emojis or the words 'push' and 'pin', and removed posts that were made outside the period of study (before June 2019 when the protest started). After these processes, 37051 posts remained. We exported all posts from the two channels (10,965 from Dadfindboy and 1,062 from Youcangotojail) into Excel spreadsheets for manual coding. We developed a coding matrix with examples. Our research assistants first coded 50 posts which were exchanged amongst the team to test inter-coder reliability. This data is sensitive, even more so since the passage of the *2020 National Security Law* (NSL). To enhance the privacy and security of the data as well as the researchers, the scraped data was stored outside of Hong Kong. Access to the data was protected by password and only granted to coders.

## *Coding matrix*

Key foci of this research are to map the citizenry's datafication of the state (and each other), the role doxing played in mobilization and to understand the individual and collective elements of doxing within the context of contentious politics. To achieve these aims, it is important to distill: 1) the type of information participants collected and released; 2) the real or intended consequences of the information release; 3) the sentiments used to frame the release – particularly those that reference a collective identity; 4) how the posts align with the persuasive functions of movement. Thus, we developed a coding matrix (see Table 1) drawn from Marx's (1999) seven types of identity knowledge and Douglas (2016) and Trottier's (2017) types and intent of doxing.

The disclosure of police and protester information in the Dadfindboy and Youcangotojail Telegram channels fits the criteria to be both de-anonymizing and targeted types of doxing, by virtue of being posted in forums established with the specific purposes of revealing the perceived opponents' identity to yield consequences (punishment). Thus, to interrogate differences in the types of doxing within this setting, we introduced an additional criterion to distinguish between posts that merely de-anonymized subjects and those that actively sought to target subjects. Posts that released information and contained some narrative calling for the harassment of or action against subjects were treated as targeting doxing.

For LIHKG data, to further explore the intent behind the doxing practices, we supplemented the three categories listed above with sentiment and content analysis in order to

**Table 1.** The coding matrix for the two Telegram channels.

| Identity knowledge type | Description |
|---|---|
| Legal name | The name under which someone is known for official and legal purposes |
| Locatability | Information that reveals where someone lives or where s/he can be contacted personally. |
| Pseudonyms (linked to name or location) | A name or code that represents a single individual such as a bank account number in a system that is related to their legal name. |
| Pattern knowledge | Someone who can be recognized by their regular public actions or habits, such as catching the same bus at the same time every day. |
| Social categorization | Information that can be used to place someone in social categories (or stereotypes) such as physical appearance, accent, style of dress, etc. |
| Symbols of eligibility/non-eligibility | Possessing artefacts or knowledge that identifies someone as being entitled to particular privileges and treatment, such as a uniform, a password etc. |
| **Intent of doxing** | |
| Harass | Include remarks that incite irritating pranks |
| Punish | Include statements/words such as: creating troubles for XXX |
| Hold to account | Include statements/words such as: sending XXX to jail or for public judgement |
| Degrade | Include statements/words such as: dog, retarded, low-educated, etc. |
| **Types of doxing** | |
| De-anonymizing | Disclosure of legal name and locatability of the subject, presented *without* any 'pattern knowledge' and *without* any narrative from the poster. |
| Targeting | Disclosure of legal name and locatability of the subject *accompanied by* details that establish 'pattern knowledge' and 'social categorization' |
| De-legitimizing | Photos of the subject showing them transgressing social norms. Accounts detailing the subjects excessive use of force or abuse of their power as a police officer. |
| **Sentiments** | |
| Patronization | Terms or expressions used to patronize and tease targets such as: wake up, daydream, face the reality, idealistic, wishful thinking, silly, poor thing, etc. |
| Sarcasm | Tones or expressions used to mock targets such as: what a brave soldier, what a handsome policeman, beautiful police wives, hard-working police, great communism, etc. |
| Anger | Expressions show one is upset and feel the current status quo is unfair, such as: pissed off, feel unfair, upset, fuck your mother, fuck you, fuck off, fuck bad dog, punishment to your family and children, etc. |
| Insults | Terms or expressions that degraded the subject in some way, such as: dog, dickface, (fucking) cunt, undercover dog, slut, rapist, cop cockroaches, blue dead bodies, low IQs, etc. |
| Cheers | Terms or expressions used to mobilize and cheer participants, such as liberate Hong Kong, revolution of our times, hang in there, brothers and sisters, thank you brothers and sisters, need people to back up, be water, etc. |

understand how doxing configured the persuasive functions of social movements. To increase the data accuracy and be more target-oriented in the sentiment analysis, we used a process of inductive coding whereby we reviewed 1,000 posts and identified regularly appearing terms and the dominant sentiments they convey. The dominant linguistic devices include: cheers, insults, curses, sarcasm and patronization.

## Findings

### Channel content and positioning

Perhaps unsurprisingly, users in both Dadfindboy and Youcangotojail channels posted content not expressly related to doxing. Dissemination of news, attitudes towards data collection and educative material appeared in similar proportions on both channels. Interesting differences in the discourse found on the two channels were also observed. As Table 2 suggests, although Dadfindboy has the reputation of being a channel dedicated to doxing, doxing only makes up 42% of its content. In contrast, the content of Youcangotojail is more concentrated on doxing (78%). The rest of the post includes sharing the news (9%), data tactics education (6%), and other physical world (non-data related) strategies such as making gas bombs or when to retreat.

**Table 2.** Post content of the two Telegram channels.

| Post content | Youcangotojail (based on 10965 posts) | Dadfindboy (based on 1,062 posts) |
|---|---|---|
| Doxing | 78% | 42% |
| News | 15% | 9% |
| Purpose of channel | 2% | 1% |
| Attitudes towards data collection | 5% | 5% |
| Education on data tactics and resistance | 1% | 6% |
| Carrie Lam[9] | 0% | 6% |
| Physical protest field strategies | 0% | 31% |
| Total | **100%** | **100%** |

We examined the linguistic devices utilized by participants across LIHKG and the two Telegram channels (see Table 3). We found that although insults and sarcasm are prevalent overall, Youcangotojail appeared to have more sarcasm and condescension, whereas in Dadfindboy and on LIHKG insults were the dominant method of expression. As Table 4 suggests, on LIHKG the comments containing insults and curses mostly targeted the police, whereas a higher proportion of sarcastic comments were directed towards China supporters and the spouses of police officers.

**Table 3.** Linguistic device on LIHKG directed at target.

| Linguistic device | Youcangotojail | Dadfindboy | LIHKG |
|---|---|---|---|
| Insults | 30% | 88% | 58.3% |
| Curses | - | - | 9.9% |
| Cheers | 1% | 1% | 13.9% |
| Patronization | 6% | 0% | - |
| Patronization + insults | 2% | 0% | - |
| Patronization + sarcasm | 2% | 0% | - |
| Sarcasm | 42% | 11% | 17.8% |
| Sarcasm + insults | 14% | 0% | - |
| Total | **100%** | **100%** | **100%** |

**Table 4.** Linguistic device of the two Telegram channels and LIHKG.

| Target | Insults | Cursing | Sarcasm | Cheers |
|---|---|---|---|---|
| Police | 66.64% | 82.7% | 20.18% | 0% |
| Police spouse | 11.35% | 0.11% | 24.08% | 0% |
| Police family members | 16.83% | 17.18% | 11.16% | 0% |
| Pro-government supporters | 5.19% | 0% | 0.65% | 100% |
| China supporters | 0% | 0% | 24.12% | 0% |
| Fifty cents/undercovers | 0% | 0% | 19.80% | 0% |
| % of total 37,051 posts | **58.3%** | **9.9%** | **17.8%** | **13.9%** |

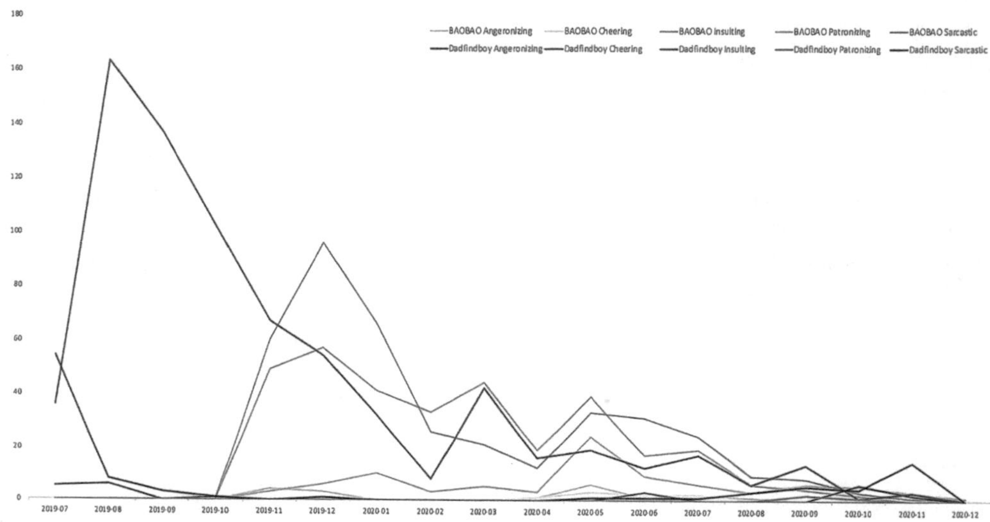


**Figure 1.** The line graph of sentiments on the two Telegram channels over time (BAOBAO here refers to Youcangotojail channel).

On Telegram, users in Youcangotojail used a more varied range of linguistic techniques than those in Dadfindboy. Interrogating the temporal dimension of the data highlighted shifts in the use of linguistic devices over time (see Figure 1). Users in Youcangotojail cycled through four linguistic devices – sarcasm, insults, sarcasm again, and finally condescension. The shift between these techniques coincides with key phases in the protest cycle. In contrast, up until the passage of the NSL in June 2020, users in Dadfindboy consistently employed insults to express themselves, giving the appearance of a protracted protest event between November 2019-June 2020. The prevalence (and peak) of insults and threats from late September to late-October 2019 coincides with the most violent period of the movement. It is likely this linguistic departure from the pro-democracy movement's motto of being 'peaceful, rational and non-violent' (*wo-lei-fei*) is linked to the escalating episodes of violence between protesters and police. In May 2020, there is a drop in the use of insults towards police and sarcasm becomes more widely used. This change in linguistic device is probably the result of the increased legal cost of insulting police officers under the revised political opportunity structure created by the Chinese state through the NSL. The apparently coordinated use of linguistic devices by both channels also confirms the emergence of a contentious repertoire as Tilly (2006, 2008) defines it.

Protesters infiltrated Youcangotojail and posted images and videos of the attacks in Yuen Long carried out by masked men in white shirts against movement sympathizers. These posts asked users to widely share these footages to allow people around the world to know how riotous and violent these 'protesters' are. These users adopted satire to achieve the goal of balancing anti-protest discourse as well as mobilizing more people in support of protest.

### *Evolution of doxing as a contentious repertoire*

Activity on Dadfindboy peaked between July and October of 2019, the time when street protests were at their peak, and also the time between the announcement of the Extradition Bill and its withdrawal in October 2019. The datafied practice of collecting and releasing police officers' information as contentious practices matured quickly. In the beginning of Dadfindboy's existence, participants simply posted photos of individual police officers and numbered them in chronological order of appearance. Others would respond with information and quote the reference number, thus giving the impression that the number of appearance was a 'reference number' in a cataloguing system for a 'database' of officer information. After July of 2019, photos that showed police officers with family members and their social media contact details also began appearing. Posts progressively became more detailed and by the end of July a template for posting information was established. The template encouraged participants to provide as much detailed information as possible about the police officer, their spouse and their children. The creation of a template emphasized collective ownership of and responsibility for the collation and use of the information in meaningful ways. The template was often used in isolation, that is to say that users did not add their own views or commentary to the post. This had the effect of framing the post as a *pseudo* neutral, factual information release or 'database entry'. In October 2019, citing the malicious intent of the Telegram channels' organizers, the Hong Kong High Court granted an interim injunction to ban the doxing of Hong Kong police officers on Telegram LIHKG (Chan, 2019). This decision symbolized the state's denunciation of doxing.

In response to Dadfindboy, government supporters established Youcangotojail channel in November 2019. Activity on Youcangotojail channel was highest between November 2019 and January 2020, with a small spike in traffic again in May 2020 when protests against the introduction of the NSL occurred. 'Anti-protester surveillance', 'anti-violence', and 'anti-riot', were the major expressions that appeared within the channel's discourse. The data collected in this channel typically included photographs of protesters engaging in 'violent' and 'illegal activities', and the location where the photo was taken, rather than providing information based on a template. Data collection and sharing practices in this channel did not evolve to the same degree as was seen in Dadfindboy. Participants posted photos of people attending rallies and captioned the photo with the date, time and place the photo was taken. These practices combined created the impression of an alliance between users and the police. Users framed themselves as releasing information about illegal activities that police could follow up and take action on. This 'adopted' proximity to state power may have influenced the connective logic among users on Youcangotojail channel. With an assumption that the state has 'better' surveillance capacity and technology, detailed templates for data input become unnecessary.

Supporting Tilly's (2006, 2008) conceptualization of contentious repertoire, the performances exchanged between the claimant-object pairing evolved dialectically and ultimately to shifts in the political opportunity structure. The contentious repertoire based in the digital world influenced and shifted the course of the contentious repertoire manifest in the physical world. The exposure of police officer's personal information led them to stop wearing identification badges and start wearing masks to cover their faces (Mozur, 2019).

### *Targets, types and intent of doxing*

Three claimant-object pairings were observed across LIHKG and the two Telegram channels. In Youcangotojail, government supporters positioned themselves as claimants against movement supporters. In Dadfindboy and on LIHKG, the police and their families were targeted by movement supporters. After the passage of the NSL a third claimant-object pairing emerged on LIHKG – movement participants and individuals suspected of being members of the fifty cents army, who are thought to be surveilling and reporting the protesters. Users were alive to the possibility that their networks had been infiltrated by government supporters and police, and discourse participants engaged in surveillance of the posted content. Any posts that seemed suspicious were flagged and identifying information about the poster was collected and publicized. A google document titled 'the user accounts of fifty cents who posted fake news on LIHKG' was created. The information shared includes the user IDs of participants believed to be members of the fifty cents army. To substantiate the assertions, the online behaviors of the individual were documented (such as commenting in support of greater unification between Hong Kong and China, remarks against independence, and spreading fake news), and links to the individual's social media profiles were included. Going beyond these obvious differences in doxing targets, an interesting variation was observed between the two channels. During the period of study, 65% of data releases in Dadfindboy targeted individual officers. The remaining 35% of releases related to the officers' family members. In Youcangotojail channel, a different pattern is seen —67% of data releases relate to groups of people and only 33% focus on a specific

**Table 5.** Types of doxing on two TG channels.

| Intent of doxing | Youcangotojail | Dadfindboy |
|---|---|---|
| Degrade | 8% | 9% |
| Harass | 0% | 1% |
| Justify doxing | 47% | 5% |
| Justify doxing and harass | 42% | 0% |
| Justify doxing and punish | 1% | 0% |
| Punish | 0% | 1% |
| Threaten | 0% | 83% |
| Total | **100%** | **100%** |

individual. The difference in intended use of the data (to assist police identify protesters) is a likely explanation for this (see Table 5).

Doxing that targeted its subjects was the dominant type of doxing seen on Youcangotojail and Dadfindboy (see Table 6). In Dadfindboy, the identity knowledge types most released included officer's legal name, locatability and pseudonyms such as badge numbers and car number plates. However, posts in Youcangotojail channel rarely released identity knowledge of the types Marx's (1999) set out. Instead, Youcangotojail channel's users practice of posting photographs containing a few protesters' faces suggests an assumption that in the age of advanced facial recognition software, an individual's face can now be an identity knowledge type that can be used to find out an individual's legal name and locatability.

**Table 6.** Intent of doxing in two TG channels.

| Type of doxing | Youcangotojail | Dadfindboy |
|---|---|---|
| De-anonymizing | 1% | 1% |
| De-legitimizing | 0% | 3% |
| Targeting | 82% | 96% |
| Targeting and de-legitimizing | 17% | 0% |
| Total | **100%** | **100%** |

The Telegram channels played the role of coordinating actions that go beyond discussion on social media. Participants came up with practices that used the private data of police as a weapon to instigate the harassment of these individuals. Examples included putting the addresses and phone numbers of police officers on to flyers advertising the sale or rent of apartments, school enrollment and restaurant deals. This practice thus drew in unsuspecting civilians as those interested in buying or renting an apartment were co-opted into the harassment of police officers. Other strategies focused on generating anxiety of the unknown, for example one post stated:

> Extending punishment to their wives and children is not just about doxing per se, we want to create a scenario in which the police will be fearful of their children being bullied on campus, or having something 'extra' in their meals when eating in a restaurant . . .

By justifying the doxing of wives and children as parties linked to but distinct from the primary target of the doxing – police officers, this comment suggests an awareness that doxing in certain circumstances might be denounced, which means people might find it morally questionable and therefore an unacceptable tactic.

### *Justification for doxing*

In Dadfindboy, the justifications given for doxing typically fall into one of two categories. That the police 'deserve' the treatment, or that the threat to Hong Kong's democratic freedoms necessitates extreme measures like doxing. In the former line of reasoning, police are framed as agents of Chinese state power, excessively violent and/or as having questionable morals. Examples include: '#XXX, has a history of stealing and secretly filming ladies' underwear, beware of this pervert everyone!'; 'a dog has to be responsible for its misbehavior, it's not innocent!' and 'so many people and their parents and friends are hit by the police, don't they [the police] need to pay the price for the violence?' In the latter, the power disparity between the people and the state is highlighted, so too is the importance of movement aims such as democracy. Comments in this category are in closer alignment with the commonly cited aims of the movement. Examples include: 'We can't fight against the police, they have powerful weapons, but we can create influence in their everyday lives . . . '

In Youcangotojail government supporters justify doxing with reference to 'maintaining social stability.' Protesters are often framed as 'rioters' or 'cockroaches'—a slur implying they are subhuman but resilient – which provides the rationale for harsh and cruel treatment against protesters. Photographs of protesters are often captioned with descriptions such as 'sneaky cockroaches hanging around the streets, must be doing something bad'; ' . . . a group of rioters used hammers and sticks to attack the police and destroy shops. It's nowhere close their claims of being peaceful, rational, and non-violent'; 'there will and must be karma for the rioters who doxed the police; you followed the police car and then we follow you!' Posts were also accompanied by remarks such as 'remember you can go to jail' or 'you dare to protest in the street then I dare to take photos of you!'

## Discussion

The 2019 Hong Kong protest has substantiated that data practices are of increasing importance in social movements' strategies as well as a debated subject. Empirically, this article has documented the creative incorporations of data-based practice into Hong Kong's contentious repertoire. Through data practice and perception of it, the action repertoire changed and new forms of mobilization/conflict also emerged (Flesher Fominaya & Gillan, 2017; van der Graaf et al., 2016). Unlike the #MeToo movement in China, India, and Turkey (Bitso et al., 2012), in the Hong Kong case we saw two ideologically opposed groups weaponizing data against each other in the same way (doxing), and the same data objects (legal name, locatability, and pattern knowledge) becoming the source of contention. However, the justification of these practices was very different. In the situation where opposing camps have weaponized data in the same way, and the same data objects are the subject of contention, the different moral justifications of doxing cause us to rethink the positionality of data as weapon vis-a-vis data as object. This extends Beraldo and Milan's (2019) observation of the coinciding aspects of data by showing that the relationship between these components does not just exist in a constant state of contradiction as the means and the object of contention. Instead, the weaponization (means) and data objects (ends) can be aligned and realigned with each other in a form of symbiosis where means and ends are used to mutually justify the doxers' ultimate end – the de-legitimization of their opponent.

### *Datafication and power*

The state's datafication of individuals through the collection of legal names, locateability and pattern knowledge was a technique refracted through participants and shone back onto the state. When used by protesters, the process of datafication enabled the various levels of identity of police officers to be brought together. Datafication in this instance positioned officers as members of a whole community, with a name, address and family members rather than as individuals identified by numbers. This has important consequences for power dynamics. De-anonymizing the police officers (e.g. disclosing their badge numbers) and casting them as individuals, as citizens, strips them of their powers as police and the protection of their government role. Instead, it places them on par with protesters. This removal of the role, in other places, has given police officers and other government officials the opportunity to take action as individuals (as we saw in Black Lives Matter protests where policemen supported protesters). However, in this case, removing the distinction between the officers' private and public lives has resulted in significant damage at the personal level.

By the same token, the datafication of the protesters by government supporters had the effect of changing their social status. Collecting footage of other citizens positioned government supporters as *proxies* of the state, sort of 'watchmen'. Then, through the process of sharing the information publicly (and thus with the state) individual participants became *agents* of the state. Further, the practice of posting photos of protesters captioned with the misdeeds they are alleged to have committed removes their identity as citizens standing up for their rights and the legitimacy that is afforded by such identity. The photos become a contemporary 'wanted poster' framing protesters as criminals who will be sought by the authorities.

### *Data is weaponized for 'veillance'*

Collecting and revealing personal data has become an important individually executed, yet collective strategy that can be used to challenge or reinforce existing power asymmetries. At one level, these findings highlight the individualized nature of doxing as a contentious performance that originates and is implemented in an individual's daily life through personal frames. Individuals post the personal information of others, in separate but connected (by virtue of being co-located in a Telegram channel) acts of doxing. The desired outcomes of these acts of doxing also lie at the individual level. Participants in Dadfindboy seek to increase the police's fears and disrupt or damage the personal lives of police officers and their family members, whereas Youcangotojail aims to trigger legal charges against protesters as private citizens. When viewed as part of a coordinated and collaborative campaign, the individual instances of doxing can be seen as coordinated programs of sousveillance and surveillance in the respective cases of Dadfindboy and Youcangotojail channel. The concentration of doxing, and therefore data, in two locations had the effect of converting the data into a database that was searchable and could be manipulated if additional steps were taken. The data practices in both instances were consistent with surveillance techniques employed by the state. Thus, our findings drawn from LIHKG and the two Telegram channels suggest that the doxing in the Hong Kong example became forms of sousveillance and surveillance that blended more traditional image-based monitoring techniques with more contemporary techniques of datafication.

## Conclusion

Situating doxing within the ontologies of data activism comes up against the norms of data release in other contexts, and challenges our understanding of what ethical data practices are and when information releases are justifiable. Our finding that doxing can be a form of sousveillance or surveillance has the effect of conflating several distinct processes each with their own moral codices. Doxing is typically associated with malicious intent and as an unethical practice in light of the damage that can be done to targets. Surveillance has both positive and negative connotations whilst sousveillance is generally understood to be an acceptable practice motivated by a sense of justice. The ethical grey zone in which doxing stands, as a form of data-activism, could pose limits to its use as a protest tactic and negatively influence the legitimacy of movement participants and their claims. The justifiability of doxing as a form of data-activism requires further research. Specific attention should be given to determining the proportion of individuals whose data was released that experienced damage, the nature of the damage experienced, and the perceptions the general public hold of doxing.

## Notes

1. See: https://www.bayarea.gov.hk/filemanager/en/share/pdf/Outline_Development_Plan.pdf.
2. The Office of the Privacy Commissioner for Personal Data (PCPD) defines 'personal data' as 'information which relates to a living individual and can be used to identify that individual.' See: https://www.pcpd.org.hk/english/data_privacy_law/ordinance_at_a_Glance/ordinance.html.
3. Millions took to the streets of Hong Kong in protest against a proposed amendment to the Extradition Law, that would permit the extradition of individuals within Hong Kong's jurisdiction to China.
4. The online forum LIHKG is widely regarded as a central communication platform for Anti-ELAB movement supporters (Lee, 2020).
5. See: https://t.me/s/Dadfindboy. It is one of the leading channels created in response to police officers not wearing identification badges. It then evolved to even advocate for harassment and violence against police officers (Maragkou, 2019).
6. See: https://t.me/s/canigosomeplacefun.
7. 'Fifty cents army' is responsible for uncovering the state's extensive efforts to delude negative content, and fostering a more legitimate vision of the party (Repnikova & Fang, 2018).
8. See: https://www.knime.com/.
9. Carrie Lam is the current Chief Executive of Hong Kong since 2017.

## Acknowledgements

We would like to thank Shao-Fang (Christine) Kao for scraping the data from LIKHG and Mei Ling (Jay) Chan for data analysis. We appreciate the help of Ruiheng (Peter) Li and Lok Hei (Ken) Liu with manually coding the data from the two Telegram channels. The draft was presented at the paper development workshop organized by Stefania Milan and Davide Beraldo, and work-in-progress workshop by the Society for Hong Kong Studies (SHKS). We appreciate the valuable feedback from the participants, especially Vassilis Charitsis, Cristina Flesher Fominaya, Stefania Milan, Davide Beraldo, Francis Lee, Ching Kwan Lee, and Edmund W. Cheng. We extend our thanks to Samson Yuen, two anonymous reviewers, and the guest editors for their constructive comments and suggestions.

## Disclosure statement

No potential conflict of interest was reported by the authors.

## Funding

The work was supported by the Strategic Priority Funding scheme, Arts, Design and Architecture (ADA), University of New South Wales [SPF02].

## ORCID

Yao-Tai Li http://orcid.org/0000-0002-4897-3616

## References


Benford, R., & Snow, D. (2000). Framing processes and social movements: An overview and assessment. *Annual Review of Sociology*, *26*, 611–639. https://doi.org/10.1146/annurev.soc.26.1.611

Beraldo, D., & Milan, S. (2019). From data politics to the contentious politics of data. *Big Data & Society*, *6*(2), 1–11. https://doi.org/10.1177/2053951719885967

Bitso, C., Fourie, I., & Bothma, T. (2012). *Trends in Transition from Classical Censorship to Internet Censorship: Selected Country Overviews*. The International Federation of Library Association (IFLA). https://www.ifla.org/publications/trends-in-transition-from-classical-censorship-to-intenet-censorship-selected-country-o

Carty, V., & Reynoso Barron, F. G. (2019). Social movements and new technology: The dynamics of cyber activism in the digital age. In B. Berberoglu (Ed.), *The Palgrave Handbook of Social Movements, Revolution, and Social Transformation* (pp. 373–397). Springer.

Chan, H. (2019). Hong Kong court grants temporary injunction against inciting violence via messaging app Telegram and LIHKG forum. Retrieved from *Hong Kong Free Press*: https://hongkongfp.com/2019/10/31/hong-kong-court-grants-temporary-injunction-inciting-violence-via-messaging-app-telegram-lihkg-forum/

della Porta, D., Peterson, A., & Reiter, H. (Eds.). (2006). *The policing of transnational protest*. Ashgate.

Dencik, L., Hintz, A., & Cable, J. (2016). Towards data justice? The ambiguity of anti-surveillance resistance in political activism. *Big Data & Society*, *3*(2), 1–12. https://doi.org/10.1177/2053951716679678

Douglas, D. M. (2016). Doxing: A conceptual analysis. *Ethics and Information Technology*, *18*(3), 199–210. https://doi.org/10.1007/s10676-016-9406-0

Earl, J., & Soule, S. (2006). Seeing blue: A police-centered explanation of protest policing. *Mobilization: An International Quarterly*, *11*(2), 145–164. https://doi.org/10.17813/maiq.11.2.u1wj8w41n301627u

Fish, A., & Follis, L. (2016). Gagged and doxed: Hacktivism's self-incrimination complex. *International Journal of Communication*, *10*, 3281–3300.

Flesher Fominaya, C., & Gillan, K. (2017). Navigating the technology-media-movements complex. *Social Movement Studies*, *16*(4), 383–402. https://doi.org/10.1080/14742837.2017.1338943

Foucault, M. (1977). *Discipline and Punish: The birth of the prison*. Penguin.

Haggerty, K. D., & Ericson, R. V. (2000). The surveillance assemblage. *The British Journal of Sociology*, *51*(4), 605–622. https://doi.org/10.1080/00071310020015280

Hermida, A., & Hernández-Santaolalla, V. (2018). Twitter and video activism as tools for counter-surveillance: The case of social protests in Spain. *Information, Communication & Society*, *21*(3), 416–433. https://doi.org/10.1080/1369118X.2017.1284880

Hoyng, R. (2021). From open data to "grounded openness": Recursive politics and postcolonial struggle in Hong Kong. *Television & New Media*, *22*(6), 703–720. https://doi.org/10.1177/1527476420931444

Jackson, W., Gilmore, J., & Monk, H. (2019). Policing unacceptable protest in England and Wales: A case study of the policing of anti-fracking protests. *Critical Social Policy*, *39*(1), 23–43. https://doi.org/10.1177/0261018317753087

Kazansky, B., Torres, G., van der Velden, L., Wissenbach, K. R., & Milan, S. (2019). Data for the social good: Toward a data-activist research agenda. In A. Daly & M. Mann (Eds.), *Good data* (pp. 244–259). Institute of Network Cultures.

Lee, F. (2020). Solidarity in the Anti-Extradition Bill movement in Hong Kong. *Critical Asian Studies*, *52*(1), 18–32. https://doi.org/10.1080/14672715.2020.1700629

Lehtiniemi, T., & Haapoja, J. (2020). Data agency at stake: MyData activism and alternative frames of equal participation. *New Media & Society*, *22*(1), 87–104. https://doi.org/10.1177/1461444819861955

Lyon, D. (2018). *The culture of surveillance: Watching as a way of life*. Polity Press.

MacAllister, J. M. (2017). The doxing dilemma: Seeking a remedy for the malicious publication of personal information. *Fordham Law Review*, *85*(5), 2451–2483.

Mann, S. (2005). Sousveillance and Cyborglogs: A 30-Year empirical voyage through ethical, legal, and policy issues. *Presence Teleoperators and Virtual Environments*, *14*(6), 625–646. https://doi.org/10.1162/105474605775196571

Mann, S. (2013). Veillance and reciprocal transparency: Surveillance versus sousveillance, ARGlass, lifeglogging, and wearable computing. Available at: http://wearcam.org/veillance/veillance.pdf

Mann, S., & Ferenbok, J. (2013). New Media and the power politics of sousveillance in a surveillance-dominated world. *Surveillance & Society*, *11*(1/2), 18–34. https://doi.org/10.24908/ss.v11i1/2.4456

Maragkou, E. (2019). "Dadfindboy": How activists in Hong Kong are hijacking state tools of surveillance. Retrieved from: http://mastersofmedia.hum.uva.nl/blog/2019/09/23/dadfindboy-how-activists-in-hong-kong-are-hijacking-state-tools-of-surveillance/

Marx, G. T. (1999). What's in a name? Some reflections on the sociology of anonymity. *The Information Society*, *15*(2), 99–112. https://doi.org/10.1080/019722499128565

McNealy, J. (2018). What is doxxing, and why is it so scary? Retrieved from *The Conversation*: https://theconversation.com/what-is-doxxing-and-why-is-it-so-scary-95848

Meng, A., DiSalvo, C., Tsui, L., & Best, M. (2019). The social impact of open government data in Hong Kong: Umbrella Movement protests and adversarial politics. *The Information Society*, *35*(4), 216–228. https://doi.org/10.1080/01972243.2019.1613464

Milan, S. (2015). From social movements to cloud protesting: The evolution of collective identity. *Information, Communication & Society*, *18*(8), 887–900. https://doi.org/10.1080/1369118X.2015.1043135

Milan, S., & van der Velden, L. (2016). The alternative epistemologies of data activism. *Digital Culture & Society*, *2*(2), 57–74. https://doi.org/10.14361/dcs-2016-0205

Mozur, P. (2019). In Hong Kong Protests, Faces Become Weapons. Retrieved from *The New York Times*: https://www.nytimes.com/2019/07/26/technology/hong-kong-protests-facial-recognition-surveillance.html

Pellegrino, G., Söderberg, J., & Milan, S. (2019). Datafication from below: Epistemology, ambivalences, challenges. *Tecnoscienza: Italian Journal of Science and Technology Studies*, *10*(1), 89–113.

Repnikova, M., & Fang, K. (2018). Authoritarian participatory persuasion 2.0: Netizens as thought work collaborators in China. *Journal of Contemporary China*, *27*(113), 763–779. https://doi.org/10.1080/10670564.2018.1458063

Schlembach, R. (2018). Undercover policing and the spectre of 'domestic extremism': The covert surveillance of environmental activism in Britain. *Social Movement Studies*, *17*(5), 491–506. https://doi.org/10.1080/14742837.2018.1480934

Schrock, A. R. (2016). Civic hacking as data activism and advocacy: A history from publicity to open government data. *New Media & Society*, *18*(4), 581–599. https://doi.org/10.1177/1461444816629469

Stewart, C. J., Smith, C. A., & Denton, R. E. (2007). *Persuasion and social movements*. Waveland Press.

Stoycheff, E., Liu, J., Xu, K., & Wibowo, K. (2019). Privacy and the Panopticon: Online mass surveillance's deterrence and chilling effects. *New Media & Society*, *21*(3), 602–619. https://doi.org/10.1177/1461444818801317

Tarrow, S. (2011). *Power in Movement: Social Movements and Contentious Politics*. Cambridge University Press.

Tilly, C. (2006). *Regimes and repertoires*. University of Chicago Press.

Tilly, C. (2008). *Contentious performances*. Cambridge University Press.

Trottier, D. (2017). Digital vigilantism as weaponisation of visibility. *Philosophy & Technology*, *30* (1), 55–72. https://doi.org/10.1007/s13347-016-0216-4

Tufekci, Z. (2017). *Twitter and Tear Gas: The Power and Fragility of Networked Protest*. Yale University Press.

Ullrich, P., & Knopp, P. (2018). Protesters' reactions to video surveillance of demonstrations: Counter-moves, security cultures, and the spiral of surveillance and counter-surveillance. *Surveillance & Society*, *16*(2), 183–202. https://doi.org/10.24908/ss.v16i2.6823

Ullrich, P., & Wollinger, G. R. (2011). A surveillance studies perspective on protest policing: The case of video surveillance of demonstrations in Germany. *Interface: A Journal for and About Social Movements*, *3*(1), 12–38.

van der Graaf, A., Otjes, S., & Rasmussen, A. (2016). Weapon of the weak? The social media landscape of interest groups. *European Journal of Communication*, *31*(2), 120–135. https://doi.org/10.1177/0267323115612210

Wang, P., Joosse, P., & Cho, L. L. (2020). The evolution of protest policing in a hybrid regime. *The British Journal of Criminology*, *60*(6), 1523–1546. https://doi.org/10.1093/bjc/azaa040

Wood, M. A., & Thompson, C. (2018). Crowdsourced countersurveillance: A countersurveillant assemblage? *Surveillance & Society*, *16*(1), 20–38. https://doi.org/10.24908/ss.v16i1.6402

# Datafication and implicated networks of demobilization: social movement demobilization in datafied societies

Chi Kwok and Ngai Keung Chan

**ABSTRACT**

An increasing number of social movements have taken the form of connective action achieved through datafication structures in the absence of central coordinating social movement organizations. Social movement demobilization in digitally-mediated connective action remains a relatively understudied area. Drawing together social movement studies and communication studies, we introduce the theoretical framework of 'implicated networks of demobilization' (INsD) to explore the dynamics and implications of datafication for social movement demobilization. Based on examples from the Anti-Extradition Bill movement in Hong Kong and the 2020 George Floyd protests in the United States, we theorize that the connective action embedded in the datafication structure confronts three demobilizational dynamics: implicated data, materiality, and epistemic dynamics. These dynamics imply that connective action mobilization in datafied societies faces constraints pertaining to state-level network oppression, potential threats imposed by the material datafication structure, and the spread of digital knowledge of self-protection. The INsD framework contributes to understanding the interactive data-mediated dynamics between the state and activists in connective action.

## Introduction

Bennett and Segerberg (2012) suggested that the rise of the digital infrastructure, especially the emergence of digitally-mediated communicative practices, has resulted in the rise of the logic of connective action, in which the centrality of social movement organizations (SMOs) declines, and personalized frames become increasingly important. Scholars have since debated the changing power dynamics in social movements resulting from expanding processes of datafication. Datafication refers to the processes through which contemporary societies become increasingly mediated by and transmuted into data (Beraldo & Milan, 2019; van Dijck, 2014). Theoretical frameworks have been proposed to capture the intersections of political power and datafication (e.g. Beraldo & Milan, 2019; Milan, 2018; Ruppert et al., 2017). Mapping the 'contentious politics of data,' Beraldo and Milan (2019, p. 1) explicate how data become a target of contentious politics ('data as stakes') and enrich the repertoire of contention ('data as repertoires'). These are important contributions that inform the potential of creative and new resistance methods enabled by datafication, and a significant step forward to conceptualizing data politics as a field of collective claim-making and power (Van Aelst & Walgrave, 2004).

As datafication reconfigures the logic of connective action, recent years have witnessed growing public concern over the disciplinary functions of data. Journalists and activists have reminded the public about how the police (mis)use data to identify and punish protesters in the Black Lives Matter (BLM) movement in the United States, including articles with titles such as 'How to Protest Safely in the Age of Surveillance' (Greenberg & Newman, 2020) and 'Going to Protest? Here's How to Protect Your Digital Privacy' (Austin, 2020). Similarly, in Hong Kong's Anti-Extradition Bill (Anti-ELAB) movement – a decentralized and networked social movement (Lee et al., 2019; Ting, 2020)—police and pro-government groups have relied on data to monitor and dox protesters (Smith, 2019).

Such surveillance potential of digital technologies speaks to the broader issue of how the expanding datafication structures, which refer to infrastructures through which 'the transformation of social action into online quantified data' occur (van Dijck, 2014, p. 198), may shape social movement demobilization, particularly in the case of state-movement dynamics in connective action that often relies heavily on datafication structures. Studies on the dynamics of social movement demobilization and datafication remain scant. Davenport (2015) argued that social movement demobilization is a result of the dynamics of outside factors (e.g. resource deprivation and state repression) and inside factors (e.g. burnout and organizational rigidity). Although his framework concentrates on the state-led demobilizing effects on SMOs, it pays relatively little attention to the effects of the rising datafication structure on repression dynamics. Classical and contemporary accounts have focused on resource mobilization as the key to explaining movement mobilization and demobilization (Davenport, 2015; Jenkins & Eckert, 1986; McCarthy & Zald, 1977; Tarrow, 2011). Demobilization occurs when states can effectively deplete the resources of SMOs (Giugni, 1998; Opp & Roehl, 1990; Tilly & Tarrow, 2015). This view focuses on repression and counter-repression dynamics between the state and SMOs (e.g. Davenport, 2015; Kalyvas, 2006).

Pinpointing the rise of 'organizationally-enabled' and 'self-organizing' networks of connective action, Bennett and Segerberg (2012, p. 756) contended that large-scale social movement mobilization can be achieved and sustained through different digital platforms in the absence of strong coordination by dominant social movement organizations. Indeed, Tilly's (1978) concept of the political opportunity structure continues to be an important tool for theorizing structural conditions for mobilization and demobilization, particularly the threats and opportunities faced by participants because of repression or facilitation by authorities (Meyer & Minkoff, 2004; Tarrow, 2011). Tarrow's concept of the 'cycles of contention' helps to identify major processes of movements, including diffusion, exhaustion, radicalization/institutionalization, and restabilization (2011, p. 198; see also McAdam et al., 2001). Relevant aspects of mobilization and demobilization include cultural framing processes, underlying social bases and values, political opportunities channeling mobilization, and supporting resources (Giugni & Grasso, 2015). Nonetheless, these frameworks have not been explicitly connected to the dynamics resulting from the ever-growing datafication structures. As such, datafication structures play an increasingly important role in mobilizational and demobilizational dynamics in emerging connective action because digitally-mediated networks perform a crucial function in organizing and sustaining movements in the absence of coordination by dominant social movement organizations. Hence, threats and opportunities that are particularly salient in datafication structures become more prominent in state–activist interactions.

Leveraging social movement studies and communication studies, we introduce the framework of 'implicated networks of demobilization' (INsD) to capture demobilization dynamics in connective action mediated by the datafication structure. Implicated networks refer to interpersonal connections built through datafication structures. We illustrate the framework with empirical examples from two recent large-scale social movements – the 2019–2020 Anti-ELAB movement in Hong Kong, and the 2020 George Floyd protests in the United States (the Floyd protests hereafter)—by attending to the state-movement dynamics. The two cases were chosen not simply because protesters relied on digital technologies for mobilization but also because police relied on such technologies for demobilization. The Anti-ELAB movement originated from an outcry against the government's proposed changes to fugitive laws, later evolving as the largest pro-democracy movement in Hong Kong's history. The movement was characterized as a leaderless connective action in which protesters relied on digital technologies for protest mobilization and coordination (Lee et al., 2019; Ting, 2020). Grounded in a longstanding history of Black resistance against police brutality and structural racism in the US (Reny & Newman, 2021), the Floyd protests reportedly had over 15 million to 26 million participants (Buchanan et al., 2020). Protesters similarly relied upon a variety of digital technologies and social media for awareness creation and mobilization, although there were growing concerns about digital surveillance among protesters (Wade et al., 2021).

We do not argue that the datafication structure is the main reason for movement demobilization, neglecting other factors, such as crowd-control agents and street protests. We acknowledge that demobilization can take many forms and that different measures of demobilization through multiple sites and means can co-exist. Whether an attempted demobilization is successful depends on a wide range of factors, such as the severity of punishment allowed in a political context and the strength of activists' networks that are beyond unilateral control of a single party. The framework aims to capture the *dynamics* of contentious interactions between the state and protesters through datafication structures. An underlying premise is that both parties account for the opportunities and threats enabled by the structures for demobilizational purposes and usually have internalized relevant concerns into their course of actions. The INsD framework serves to identify, clarify, and detail the concrete elements of these dynamics for further analytical and research purposes. We aim to use the two movements as examples to cast datafication as sites and stakes of political struggle in connective action, rather than to offer an in-depth comparative analysis of the two movements. We center on the concept of connective action rather than other digitally-mediated movements, because it signals a context where the three dynamics of the framework – the implicated data, materiality, and epistemic dynamics – are particularly salient due to the absence of central coordinating SMOs.

This article makes three contributions. First, our conceptual framework highlights the interwovenness between datafication structures and connective action. In response to Beraldo and Milan (2019)'s 'bidirectional perspective' which depicts the relationship between data and activism as 'data-as-stakes' or 'data-as-repertoire' (p. 6), we demonstrate that the demobilizational dynamics resulting from the expanding datafication structures are not a peculiar feature of data activism, but are also found in connective action that draws heavily on the structures.[1] Second, we aim to bridge the logic of connective action with a growing interest in demobilizational dynamics. Third, we highlight how protesters are implicated with datafication structures in connective action.

In the next section, we discuss the existing literature on social movement demobilization and highlight that the role of datafication has not yet been sufficiently addressed. We then explicate the INsD framework, including the implicated data, materiality, and epistemic dynamics. The final section draws on the Anti-ELAB movement in Hong Kong and the 2020 George Floyd protests in the United States to illustrate the framework. We conclude by reflecting on the utility of the INsD framework in studies of demobilization in connective action.

## Datafication and social movement demobilization

Contemporary states draw on data infrastructures to maintain political stability, such as managing and demobilizing contentious claim-making (e.g. Milan & van der Velden, 2016). An example is the Russian Yarovaya law, which came into effect in 2016 and requires telecommunications operators to store user data and information for one year. These data could be requested by the authorities for regime repression. In 2018, 'several hundred cases were initiated due to posts, reposts, comments and likes in VKontakte,' which is one of the major social media platforms in Russia that opposition movements often utilize (Poupin, 2021). This incident exemplifies the need to reflect on the changing dynamics in contentious politics resulting from data infrastructures.

Communication researchers have studied new mobilizational opportunities generated by datafication. Beraldo and Milan (2019) have demonstrated that datafication has enriched the repertoire of contention. Specifically, datafication enables 'individual acts of rebellion and practices' (p. 3), and thereby expands the scope of agents who are capable of claim-making and reappropriating data and existing power-relations. Nonetheless, we know less about the new dynamics of movement demobilization resulting from datafication, especially in connective action, where the coordinating role of SMOs becomes less important. The relative lack of attention to demobilization is a long-standing problem in social movement studies. As Taylor stated in 1989, 'scholars generally are more interested in movements undergoing cycles of mass mobilization and have done little research on movements in decline and equilibrium' (p. 772). Two decades later, Fillieule (2013) reaffirmed that social movement scholars 'know relatively little about the mechanisms governing the decline of social movements and the varied forms of individual or collective demobilization' (p. 1), while 'the logical counterpart of the initial recruitment and mobilization processes is clearly collective demobilization and individual disengagement' (p. 1).

Studies of social movement demobilization can be classified into two types. The first type examines the depletion of resources for claim-making in social movements

(Fillieule, 2013; Klandermans, 1997; Tilly & Tarrow, 2015). The second type focuses on various forms of regime repression (Yuen & Cheng, 2017). Both groups of literature can be broadly separated into two levels: the collective level and the individual level. On the collective level, demobilization happens 'in a multiorganizational field ... of an entire social movement industry, with its formal organizations, its support networks and those involved along the way' or 'the slow collapse of a mobilization campaign, as a result of its success or, conversely, its failure' (Fillieule, 2013, p. 1). Curtis and Zurcher (1973) argued that the concept of 'multi-organizational field' has indicated that few organizations 'can operate in an interorganizational void' (p. 60). Choudry and Kapoor (2013) have also examined the 'NGOization' process resulting from social movements, such as the professionalization of dissent knowledge and the emergence of NGOs that position 'themselves as the gatekeepers between social movements and other organizations' (p. 9). The collective dimension focuses on resource depletion in the mobilizational network. However, how demobilizational dynamics within datafication structures operate remains unclear. As Grimm and Harders (2018) suggested, these studies are 'endless correlation of the total aggregate level of one output (repression) with another (protest), effectively black-boxing the interaction process itself' (p. 2).

On the individual level, some studies have focused on why participants withdraw from a movement, with three main interdependent factors identified: '*exhaustion of the rewards of involvement*, *the loss of ideological meaning*, and *the transformation of relations of sociability*' (Fillieule, 2013, p. 2, emphasis original). Fillieule (2010) proposed that a career approach helps to understand how 'at each biographical stage, the attitudes and behaviors of activists are determined by past attitudes and behaviours' (p. 1), which enables the study of political disengagement to be situated into an individual's life cycle. McAdam's (1990) classical work showed that the degree of integration into activist networks matters in determining movement withdrawal, especially in high-risk activism. Scholars have argued that the movement reward and participation costs become connected to datafication structures, particularly regarding how the state and activists utilize the structures for mobilization and counter-mobilization purposes (Dencik et al., 2016).

The second group of literature within social movement demobilization is concerned with regime repression. It tackles the issue of authorities' responses to challengers in the cycle of contention (Tarrow, 2011). Four approaches are salient (Yuen & Cheng, 2017). The first approach concerns how different regime types cause different forms of regime responses, including toleration, concession, and repression (Cai, 2010; Tilly, 1978; Yuen & Cheng, 2017). The second approach concerns various forms of regime response and their respective impacts on social movements (Lichbach, 1987; Yuen & Cheng, 2017). For example, Ellefsen (2016) has directed attention to the concept of 'judicial opportunities' in movement demobilization, that is, 'the possibilities and threats perceived and experienced by protestors' (p. 444). Such perceived threats can be creatively exploited by private elites and criminal justice agencies for legal repression against contentious groups (Ellefsen, 2016). The third approach examines variations within each type of regime response (Earl, 2003; Yuen & Cheng, 2017). The fourth approach, what Yuen and Cheng (2017) called 'regime attrition,' examines 'a mode of regime response that only tolerates protests ostensibly but uses a proactive tactical repertoire to discredit, wear out and increase the cost of protests' (p. 613). Nonetheless, this strand of the literature presupposes a relatively active and intentional agent with a clear purpose in the process of demobilization. It cannot sufficiently address the *structural* demobilizational dynamics caused by the ever-expanding scope of datafication. For example, when activists perceive that they are being targeted by the state and counter-movements, they may engage in forms of self-censorship online (Tufekci, 2017). Structural demobilizational dynamics, therefore, can trigger counter-surveillance dynamics, even without a clear repressive agent acting actively.

Tilly and Tarrow (2015) defined mobilization as 'an increase of the resources available to a political actor for collective making of claims,' whereas demobilization 'is a reduction of this aggregation of resources' (p. 120). In theory, datafication can enrich both the repertoire of contention and the repertoire of repression. Much research has focused on how creative subversive agents can appropriate data and take advantage of the ever-expanding data infrastructures. However, the state can also take advantage of the data infrastructure for demobilization by demonizing online platforms, creating information gluts, and doxxing (Tufekci, 2017).

## Implicated networks of demobilization

Given the growing impact of datafication on social movement demobilization, we introduce the INsD framework. The term 'implicated networks' highlights that the ever-expanding data infrastructure is a key networked public sphere for both mobilization and demobilization in connective action. There are three key dimensions of the INsD framework.

The first is the *implicated data dynamics*. The datafication structures have created 'implicated networks' in which protest coordination and immediate tactical communication are increasingly conducted online, and access to such information is increasingly dependent on access to different communicative networks. The interactions between actors leave relatively durable and traceable imprints for policing. Thus, the state exercises demobilizational tactics through the datafication structure. Datafication enables the state to target and monitor citizens for *reactive* and *preventive* demobilization. In reactive demobilization, the government may police a massive number of loosely connected participants by gaining access to the mobilizational network and platforms that activists rely upon for law enforcement purposes. Preventive demobilization focuses on how data can be used to sketch the networks of protesters for future policing purposes. Digital traces and relational data left by protesters can potentially be used for future demobilizational purposes or for building a network of close monitoring.

The second is *materiality dynamics*. The materiality dimension of datafication structures increasingly becomes the object of mobilization and demobilization. Material objects include mobile phones and material infrastructures. While the government often attempts to neutralize the material dimensions of the datafication structure in the name of efficiency, mobilization may occur through politicization of the material structure. This type of neutralization–politicization dynamics is connected to the discursive effects of datafication. As Benford and Snow (2000) suggested, social movement is about 'meaning work – the struggle over the production of mobilizing and countermobilizing ideas and meanings' (p. 613). Both the state and activists narrate the materiality of datafication structures differently for mobilization and demobilization.

The third is *epistemic dynamics*. Part of the determinant of the cost of social movement participation depends on the successful mastering of digital knowledge and tools to protect oneself from potential arrest. Thus, mobilization and demobilization revolve increasingly around the spread of digital knowledge concerning self-protection. The growth of datafication structures has led to an enhancing awareness that one might be subject to state surveillance (Dencik et al., 2016; Marthews & Tucker, 2017). Notably, activists attempt to spread digital knowledge concerning self-protection and anti-surveillance to incentivize participation. The adoption of privacy and identity protection technologies has become a central tactic for overcoming anticipatory surveillance.

## Situating data-mediated demobilization in contexts

### *Empirical materials*

We now turn to illustrate the framework with examples of state-movement dynamics in two large-scale social movements: the 2019–2020 Hong Kong's Anti-ELAB movement, and the 2020 Floyd protests. We selected these movements primarily because police forces reportedly deployed digital technologies and data to track down protesters who also relied heavily on digital technologies and social media for mobilization. The movements make compelling cases for considering the dynamics of demobilization in connective action mediated through the datafication structure.

We drew empirical examples from a variety of sources. In the case of the Anti-ELAB movement, we scrutinized the movement posters collected by the ANTIELAB Research Data Archive (2020).[2] The database collected such posters from two major publicly accessible Telegram channels until January 2020. The examples we discussed here were drawn from a keyword search ('surveillance'; $n = 243$) through the movement posters database. We also examined news coverage of data and surveillance in the movement published between 12 June 2019 and 30 June 2020 ($n = 99$). The news articles were collected from WiseNews. In the case of the Floyd protests, we relied upon the Electronic Frontier Foundation (EFF)'s online publications to identify instances of protest surveillance. It yielded over 20 publications published between May 2020 and September 2021. We particularly attended to how the EFF articulated the strategies that protesters could use to identify and resist protest surveillance. Wade et al. (2021) found that such documents not only identified potential counter-surveillance strategies but also revealed protesters' common concerns about digital surveillance. We also used the LexisNexis search engine to identify newspaper articles concerning surveillance in the Floyd protests published between 25 May and 31 December 2020 ($n = 66$). Lastly, we drew on existing academic studies on the two cases.

### *The Anti-ELAB movement in Hong Kong*

The Anti-ELAB movement between 2019 and 2020 was originally triggered by the government's proposed bill to establish extradition treaties with Taiwan and mainland China. Drawing on on-site protest surveys between June and August 2019, Lee et al. (2019) found that there were no recognized leaders in the movement.[3] Cheng et al. (2022) argued that the abeyance of civil society networks after the 2014 Umbrella Movement facilitated the initial mobilization and public deliberation of the movement, whereas digital communication technologies cultivated affective ties among protest groups. Protesters relied on Telegram and a Reddit-like online forum called LIHKG to coordinate and mobilize protest events (Cheng et al., 2022; Kow et al., 2020). Platform-specific affordances, Lee et al. (2021) contended, enabled movement supporters to have decentralized participation; for example, LIHKG is a large anonymous forum and does not have a network structure that connects users. Yet, LIHKG's lack of personalized popularity and credibility metrics resulted in the difficulty of sustaining opinion leadership (Liang & Lee, 2021). Telegram enabled the creation of protest-related groups and afforded movement supporters to circulate protest-related information without disclosing their personal information (Kow et al., 2020; Lee et al., 2021). The ANTIELAB Research Data Archive (2020) documented 1,097 protest events coordinated by four Telegram channels between 16 June 2019 and 27 April 2020. The movement lost its momentum after the mass arrests at university campuses in November 2019, followed by the spread of the COVID-19 pandemic and the enactment of the National Security Law in June 2020.

### *The Floyd protests in the United States*

The Floyd protests were triggered by the police killing of George Floyd in Minneapolis, Minnesota on 25 May 2020. The protests spoke to a longstanding history of Black resistance against the structural racism underlying state violence and police brutality in the US (Reny & Newman, 2021). By June 2020, more than 4,700 demonstrations had taken place in about 2,500 small towns in cities (Buchanan et al., 2020). Between 26 May and 7 June 2020, approximately 47.8 million tweets contained the #BlackLivesMatter hashtag (Anderson et al., 2020). Although the Anti-ELAB movement and the Floyd protests took place in different institutional settings and had different repertoires of contention, police reportedly used protesters' phone data and facial recognition technologies for policing and law enforcement in both cases (e.g. Schoolov, 2020; Smith, 2019).

In what follows, we will draw examples from these two cases to illustrate three core demobilizational dynamics in the INsD framework: (1) implicated data dynamics, (2) materiality dynamics, and (3) epistemic dynamics.

### *Implicated data dynamics*

The first aspect of the INsD framework concentrates on policing enabled by datafication for reactive and preventive demobilization. In both cases, there were concerns about how law enforcement officers could (mis)use protesters' data against them. In the Floyd protests, Wade et al. (2021) found that many of the publications about protest privacy recommendations explicitly articulated fears of being monitored, such as ruined reputation, online harassment, and arrest. In the Anti-ELAB movement, Kow et al. (2020) observed that there was a widespread belief among participants that 'the police and other institutional actors were monitoring and identifying the chat group' on Telegram (p. 6). The fear of surveillance was evidenced by the platform-specific affordances of LIHKG and Telegram. Both LIHKG and Telegram have features that afford movement supporters the ability to connect with others while maintaining anonymity (Kow et al., 2020; Lee et al., 2021).

There were at least two sources of data for policing – social media posts and protesters' phone data. In 2016, the Baltimore Police Department relied on real-time maps of social media activities at protest sites to identify BLM protesters (Ringrose & Ramjiee, 2020). The Los Angeles Police Department (LAPD) similarly developed a social media user guide to encourage officers to create a 'fictitious online persona' and to monitor 'content on the Internet for any discussions, posts, videos, blogs, and online conversations about the Department or other topics of interest to the Department' in 2015 (Dwyer, 2021). During the Floyd protests, the LAPD worked with Dataminr – a social media surveillance company that advertised its software as a means to monitor 'high impact events and emerging risks'—to closely observe posts regarding protest events in real time on Twitter, Facebook, Snapchat, and other social media platforms (Ray, 2021). Based on official email exchanges between the LAPD and Dataminr, the latter helped identify potential protests based on the monitoring of a public message board (Ray, 2021). Further, the Minneapolis Police Department arrested an activist for allegedly inciting riots on Facebook on 27 May 2020 (Farivar & Solon, 2020). These examples reveal how social media posts can be used for reactive and preventive demobilization.

Police could speculate whether individuals frequently participated in protests based on their phone's locational data (Ringrose & Ramjiee, 2020). Although police must obtain a search warrant to gain access to one's locational data in the US, they have increasingly opted to use geofence warrants 'which do not garner information on specified individuals but information on numerous persons within a specific area' (Ringrose & Ramjiee, 2020, p. 355). A geofence warrant may be issued when it is difficult to identify a suspect. Fussell (2021) reported that 'geofence request to Google increased from 941 in 2018 to 11,033 in 2020.' The Minneapolis Police Department, for instance, used geofence warrants to order Google to share data about any devices in the area of the AutoZone store because a masked man broke windows during a protest about Floyd's death on 27 May 2020 (Whittaker, 2021). However, protesters might not be informed about geofence warrants (Ringrose & Ramjiee, 2020). Importantly, geofence warrants enable law enforcement officers to collect data about people in the area, regardless of their participation. Such data can be used for preventive demobilization.

Similarly, the Hong Kong police force, particularly the Cyber Security and Technology Crime Bureau, actively monitored internet activities during the Anti-ELAB movement. One notable example occurred in August 2019, when a photo of a female protester who claimed she was shot in the eye was widely circulated online. The injured protester subsequently posted a short video to thank the fellow protesters for their support and to denounce police brutality. In the video, she wore a mask and used a voice changer to protect her identity. After the video was posted, the police, however, obtained a warrant to search her medical records from the Hospital Authority until the protester won a juridical review against the police's action (Smith, 2019).

Furthermore, Hong Kong police arrested several administrators of Telegram channels, allegedly for doxxing police by leaking their personal information online (Lau & Lum, 2019) and 'incit[ing] users to damage police property and interfere with traffic' (Siu & Leung, 2020), respectively, during the Anti-ELAB movement. One of the administrators was found guilty of spreading hate speech and inciting doxxing and other criminal activities via Telegram (Wong, 2021). The police could potentially screen the contact information of all these members (if the members made it available in the app) and messages in Telegram (Lau & Lum, 2019). One of those arrested – who managed a Telegram group with more than 30,000 users – stated in a media interview that he never participated in the protest; instead, he had only shared protest-related information (Pang, 2019). He also claimed that the police forced him to unlock his mobile phone and interrogated him about how the Telegram group worked. These instances reveal how platform-specific affordances could be deployed by the state to demobilize and discipline protesters.

Once protesters are arrested and their phones seized by the police, there are potential risks associated with their phone data. In 2020, the Court ruled that the Hong Kong police are authorized to 'search the digital contents of a mobile phone (or a similar device) seized on arrest without warrant only in exigent circumstances' and when 'it is not reasonably practicable to obtain a warrant' (*Sham Wing Kan v Commissioner of Police*, 2020). This ruling has expanded the police's authority to use data for demobilization. By 22 January 2020, the police arrested over 7,000 protesters (ANTIELAB Research Data Archive, 2020). The police seized and stored at least 3,700 protesters' phones at the police headquarters in Wan Chai (Ng, 2020). Then, they successfully obtained a warrant to search the headquarters to access phone data, without informing the arrestees. The police also reportedly used the phone hacking services from an Israeli firm, Cellebrite, to crack into activists' mobile devices. Although Cellebrite announced that it would stop selling the software to the police in October 2020, the police had used the software for a long time in the past. In January 2021, the police reportedly began sending the arrestees' phones to mainland China to extract the data for policing (Mahtani, 2021).

In both cases, datafication has strengthened law enforcement officers' capacity to gather protesters' digital traces and data through their implicated networks for reactive and preventive law enforcement at a distance, without the need to rely on street-level bureaucrats.

### *Materiality dynamics*

Whereas the first dimension explores how implicated networks can be used for demobilization purposes, materiality dynamics focus on how political authorities collect such data through visible and invisible data infrastructures. Protesters' mobile phones and social media posts are part of the *invisible* infrastructures because protesters do not know whether, when, and how their data could be collected. In the case of Floyd protesters, police might use cell-site simulators to identify and track their phones' international mobile subscriber IDs at protest sites (Guariglia, 2020). Only when authorities obtained warrants to search protester' phones, data-driven surveillance becomes more visible to protesters. Police officers and data analysts are often behind the scenes, but they are essential human actors for rendering data infrastructures possible, as they constantly decide what to record and whom to police (Brayne & Christin, 2021).

However, some surveillance technologies, such as CCTV cameras in public spaces, are *visible* to protesters. In the Floyd protests, the Homeland Security Department used helicopters to film demonstrations in over 15 cities (Kanno-Youngs, 2020). In San Francisco, police deployed over 400 semi-private surveillance cameras (Guariglia & Maass, 2021). Such video footage was transmitted to control centers for reactive and preventive demobilization. With the partnership with Amazon Ring and its associated app, the LAPD reportedly used a law-enforcement portal to request footage from Ring users (Guariglia & Maass, 2021). These examples show how protesters can be constantly monitored by data-driven devices.

In the Anti-ELAB movement, protest events about smart lampposts are particularly relevant to the materiality dynamics of datafication. These events offer insights into tensions between the Hong Kong government's smart city vision and protesters' counter-claims about facial recognition and data surveillance. It is noteworthy that the events took place in August 2019, before the Anti-ELAB lost its momentum. Yet, our goal is to use this example to highlight how protesters' concerns over data-driven surveillance would increase the anticipatory cost of participation.

Such tensions are revealed in the two parties' distinctive naming of the technology. The government considered the 'smart lampposts' as a key digital infrastructure project to promote cost-effective and transparent urban governance (Smart Lampposts Technical Advisory Ad Hoc Committee, 2020). By contrast, the protesters considered such a project an expanding surveillance infrastructure. A protest event called 'Light Up Hong Kong, Awake Hongkongers' was organized in Kwun Tong on 24 August 2019. The organizers' pamphlet named the lampposts as 'surveillance lampposts' with a facial recognition system. The organizers denounced that such lampposts, together with the new RFID Hong Kong identity cards, could render Hong Kong into 'a large prison' where Hongkongers 'are completely monitored.'

Protesters associated the anticipatory threats of the lampposts with those of mainland China's state-centered social credit systems. Consistent with popular surveillance imaginaries of mainland China's state-centered social credit systems (Liu, 2019), protesters contended that the government could use lamppost data to monitor people's daily activities, distribute social resources, and punish dissents. On August 21, one of the protest organizers stated that the government had not consulted the public before installing the lampposts. They were aware that the lampposts were 'only a tool,' but they did not trust 'the government behind the technology.') During the August 24 protest, some protesters used electric saws to cut down about 20 lampposts and looked into the specifications of the lampposts (Chan, 2019). They found a Bluetooth beacon called 'BLE Locator: Model SPLD01' inside such lampposts and suspected that the lampposts could be used to track people's personal and location data from their digital devices. When looking into the provider of the beacon, TickTack, protesters discovered that the provider's website was connected with Shanghai Sansi, one of the companies that supported China's digital surveillance network (Chan, 2019). As activists elaborated, although the government claimed that the lampposts were not intended to collect people's personal data, it remains unclear how the government would use the data for law enforcement.

Indeed, we can observe parallel future-oriented concerns over data misuse and surveillance in COVID-19 pandemic governance in Hong Kong. Although the government has reiterated that the contact tracing mobile app poses no privacy concerns, some citizens believe that the government is rolling out a surveillance system similar to mainland China's 'Health Code' (Li, 2021). What is at stake here is not whether the government actually collected data about protesters from surveillance cameras in the Floyd protests or the lampposts and contact tracing app in Hong Kong, but how the materiality of data-driven technologies incited protesters to anticipate surveillance and counter data surveillance accordingly, or what Kazansky (2021) termed 'anticipatory data practices' (p. 2). For protesters in the Anti-ELAB movement, wearing face masks (Mahtani, 2019) was not simply to protect themselves from the tear gas but also to prevent the threats of being identified by the materiality structures. In this vein, the introduction of surveillance technologies would increase the cost of participation.

### Epistemic dynamics

In both cases, protesters were aware of the use of data-driven technologies for surveillance and policing. Epistemic dynamics are concerned with how protesters share privacy knowledge and articulate counter-strategies against the (mis)use of data.

In the Floyd protests, civil society organizations, such as the EFF and the American Civil Liberties Union, developed a list of strategies that protesters could use to protect themselves from surveillance (Wade et al., 2021). EFF's (Electronic Frontier Foundation, 2020) guide, 'Surveillance Self-Defense,' suggested that protesters should turn on full-disk encryption on their mobile devices, remove biometric unlocking, install encrypted messaging tools, and purchase a prepaid, disposable phone before participating in protests. EFF also suggested that people should 'consider obscuring the faces of anyone who has their eyes or mouth visible' when taking and posting photos of protests on social media because the police might use such photos to identify and arrest protesters. Other common protest-related recommendations in relevant media and online publications include the use of airplane mode and the management of locational and metadata (Wade et al., 2021).

In the Anti-ELAB movement, Telegram enabled protesters to 'broadcast' and circulate protest-related information and discuss future protest actions (Kow et al., 2020). Although Telegram's anonymity features allowed users to protect their privacy, platform-specific affordances do not determine movement dynamics and outcomes (Lee et al., 2021). Instead, epistemic dynamics contributed to how digital technologies could and should be used. For instance, protesters instructed others that they should use customized privacy settings to conceal their identities (Kow et al., 2020), although they recognized that this strategy was not entirely safe. Some self-organized protest groups on Facebook further urged protesters to use a virtual private network (VPN) and an anonymous prepaid SIM card to access Telegram. The movement posters circulated through the Telegram channels also suggested that a 'safe' and 'reliable' VPN should not have any affiliations with companies from mainland China. Protesters were also instructed to avoid using public Wi-Fi in public spaces to prevent their IP addresses from being tracked. This is because the internet service providers would have phone users' location data, browsing history, and other usage data if protesters did not use a VPN and the prepaid SIM card. The government could then request such data from the service providers. These privacy instructions also encouraged protesters not to discuss their protest experiences, even with their friends. The spread of this digital privacy knowledge, together with the practices of using cash and wearing face masks during protests (Mahtani, 2019), as well as using pseudonyms on social media platforms (Smith, 2019), aimed to inform protesters to remove any digital traces that could be used by the government for repression of an individual user and their connective network.

In both cases, the opacity of data-driven technologies resulted in a dilemma – protesters lacked full knowledge about how their data could be used (Wade et al., 2021), despite having some awareness concerning the potential sensitivity of their data. While protesters' epistemic practices might lower the cost of participation, these fears testified to the idea that protesters were trapped into implicated networks. As reflected by Smith (2019), a Hong Kong-based writer who used a pseudonym to stay anonymous, regarding the Anti-ELAB movement, 'The people who built these networks [i.e. social media] didn't think about how they might work in countries where public assembly can be criminalized or a connected online identity can be grounds for arrest. Now, some protesters are realizing too late that their digital footprint is hard to erase.'

## Conclusion

Due to the rise of digital communication technologies, an increasing number of social movements have taken the form of connective action achieved through datafication structures in the absence of central coordinating SMOs. Social movement demobilization remains a relatively understudied area, let alone social movement demobilization in data-mediated connective action. Thus, we attempted to bridge communication studies with social movement studies to address this gap. We propose the INsD framework to highlight three major demobilizational dynamics – implicated data, materiality, and epistemic – which are particularly salient in connective action achieved through datafication structures.

In the Anti-ELAB movement and the Floyd protests, datafication structures became one of the major sites of struggle between the state and protesters. Concerning implicated data dynamics, accessing, tracking, and policing protesters and their implicated networks through the digital imprints that they have left is a key strategy for demobilization. In the Floyd protests, the police relied on this strategy for arresting activists and identifying potential protests. In the Anti-ELAB movement, police used protesters' phone data for law enforcement. There was also the criminalization of doxxing against the police. Regarding materiality dynamics, data infrastructures could be deployed for policing purposes. Although protesters were aware of and anticipated the surveillance threats of such infrastructures, they could not escape from data surveillance, which in turn increased the cost of participation in connective action. The epistemic dynamics are concerned with both the lack of transparency in how law enforcement officers police through data and in protesters' digital knowledge. Although the affordances of digital technologies facilitate movement mobilization (Lee et al., 2021), people's knowledge – both of counter-surveillance strategies and digital technologies – shape their anticipatory practices. The circulation of counter-surveillance strategies is evidenced in the Floyd protests and the Anti-ELAB movement. As such, epistemic dynamics illustrate protesters' efforts in hiding themselves from these surveillance threats. The article's contributions lie in articulating these three dynamics in connective action achieved through datafication structures and illustrating these dynamics by concrete cases in the Anti-ELAB movement and the Floyd protests.

The article highlights that understanding the mobilizational logic of connective action executed through datafication structures requires an understanding of the demobilizational logic of the state. Some mobilizational measures are indeed counter-demobilization measures, such as the smart lampposts in the Anti-ELAB movement and police surveillance in the Floyd protests. The INsD framework thereby helps to analyze the interactive dynamics between the state and activists that are crucial for understanding the two parties' data-mediated struggles.

Lastly, the INsD framework applies to other, more general, digitally-mediated movements and contestations between the state and protesters. An example is the outbreak of the COVID pandemic, which sparked a new wave of concern over the normalization of data collection and the potential threats to individual privacy. As Deibert (2020) put this, 'the more we embrace and habituate to these new applications, the deeper their tentacles reach into our everyday lives and the harder it will be to walk it all back.' Controversies have also been surrounding issues pertaining to the three dynamics of the INsD framework: first, about the implicated data generated by the extensive adoption of vaccine passports; second, about the monitoring infrastructures built in the name of public health; and third, about the lack of knowledge over how pandemic-related data would be used in the future. Thus, the INsD framework could be a useful tool for future research on digitally-mediated struggles.

## Notes

1. It is noteworthy that whether the dynamics exist and become prevalent depend on contextual factors such as whether there exists a well-developed datafication structure in which both the state and activists could draw on to suppress or sustain claim-makings.
2. The "ANTIELAB Research Data Archive" is an open access database created by the Journalism and Media Studies Centre at the University of Hong Kong. It extracted and stored data about "Mobilization Map," "Teargas Map," and the movement posters from the movement-related Telegram channels. The dataset, however, is now set to private. For a description of the data archive, see https://antielabdata.jmsc.hku.hk/about-us/
3. Lee et al. (2019) made the survey data about the demographics of protesters available online.

## Disclosure statement

No potential conflict of interest was reported by the author(s).

## ORCID

Chi Kwok http://orcid.org/0000-0003-3754-2447
Ngai Keung Chan http://orcid.org/0000-0002-5848-3098

## References

Anderson, M., Barthel, M., Perrin, A., & Vogels, E. A. (2020, June 10). *#BlackLivesMatter surges on Twitter after George Floyd's death*. Pew Research Center. https://www.pewresearch.org/fact-tank/2020/06/10/blacklivesmatter-surges-on-twitter-after-george-floyds-death/

ANTIELAB Research Data Archive. (2020). Retrieved 14 January 2021, from https://antielabdata.jmsc.hku.hk/

Austin, P. L. (2020, June 17). Going to protest? Here's how to protect your digital privacy. *Time*. https://time.com/5852009/protest-digital-privacy/

Benford, R. D., & Snow, D. A. (2000). Framing processes and social movements: An overview and assessment. *Annual Review of Sociology*, *26*(1), 611–639. https://doi.org/10.1146/annurev.soc.26.1.611

Bennett, W. L., & Segerberg, A. (2012). The logic of connective action: Digital media and the personalization of contentious politics. *Information, Communication & Society*, *15*(5), 739–768. https://doi.org/10.1080/1369118X.2012.670661

Beraldo, D., & Milan, S. (2019). From data politics to the contentious politics of data. *Big Data & Society*, *6*(2), 205395171988596. https://doi.org/10.1177/2053951719885967

Brayne, S., & Christin, A. (2021). Technologies of crime prediction: The reception of algorithms in policing and criminal courts. *Social Problems*, *68*(3), 608–624. https://doi.org/10.1093/socpro/spaa004

Buchanan, L., Bui, Q., & Patel, J. K. (2020), July 3. Black lives matter may be the largest movement in U.S. history. *The New York Times*. https://www.nytimes.com/interactive/2020/07/03/us/george-floyd-protests-crowd-size.html

Cai, Y. (2010). *Collective resistance in China*. Stanford University Press.

Chan, H. (2019, August 26). Hong Kong tech firm pulls out of smart lamppost programme after surveillance accusations and staff threats. *Hong Kong Free Press*. https://hongkongfp.com/2019/08/26/hong-kong-tech-firm-pulls-smart-lamppost-programme-surveillance-accusations-staff-threats/

Cheng, E. W., Lee, F. L. F., Yuen, S., & Tang, G. (2022). Total mobilization from below: Hong Kong's freedom summer. *The China Quarterly*, 1–31. https://doi.org/10.1017/S0305741022000236

Choudry, A., & Kapoor, D. (Eds.). (2013). *NGOization: Complicity, contradictions and prospects*. Bloomsbury.

Curtis, R. L., & Zurcher, L. A. (1973). Stable resources of protest movements: The multi-organizational field. *Social Forces*, *52*(1), 53–61. https://doi.org/10.2307/2576423

Davenport, C. (2015). *How social movements die: Repression and demobilization of the republic of New Africa*. Cambridge University Press.

Deibert, R. J. (2020, November 23). We've become dependent on a technological ecosystem that is highly invasive and prone to serial abuse. *The Globe and Mail*. https://www.theglobeandmail.com/opinion/article-the-pandemic-has-made-us-even-more-dependent-on-a-highly-invasive/

Dencik, L., Hintz, A., & Cable, J. (2016). Towards data justice? The ambiguity of anti-surveillance resistance in political activism. *Big Data & Society*, *3*(2), 205395171667967. https://doi.org/10.1177/2053951716679678

Dwyer, M. P. (2021). *LPAD documents reveal use of social media monitoring tools*. Brennan Center for Justice. https://www.brennancenter.org/our-work/analysis-opinion/lapd-documents-reveal-use-social-media-monitoring-tools

Earl, J. (2003). Tanks, tear gas, and taxes: Toward a theory of movement repression. *Sociological Theory*, *21*(1), 44–68. https://doi.org/10.1111/1467-9558.00175

Electronic Frontier Foundation. (2020). *Surveillance self-defense*. https://ssd.eff.org/en

Ellefsen, R. (2016). Judicial opportunities and the death of SHAC: Legal repression along a cycle of contention. *Social Movement Studies*, *15*(5), 441–456. https://doi.org/10.1080/14742837.2016.1185360

Farivar, C., & Solon, O. (2020, June 20). FBI trawled Facebook to arrest protesters for inciting riots, court records show. *NBC News*. https://www.nbcnews.com/tech/social-media/federal-agents-monitored-facebook-arrest-protesters-inciting-riots-court-records-n1231531

Fillieule O. (2010). Some elements of an interactionist approach to political disengagement. *Social Movement Studies*, *9*(1), 1–15. 10.1080/14742830903442436

Fillieule, O. (2013). Demobilization. *The Wiley-blackwell Encyclopedia of Social and Political Movements*. https://doi.org/10.1002/9780470674871.wbespm064

Fussell, S. (2021, August 27). An explosion in geofence warrants threatens privacy across the US. *Wired*. https://www.wired.com/story/geofence-warrants-google/

Giugni, M. G. (1998). Was it worth the effort? The outcomes and consequences of social movements. *Annual Review of Sociology*, *24*(1), 371–393. https://doi.org/10.1146/annurev.soc.24.1.371

Giugni, M., & Grasso, M. T. (2015). Environmental movements in advanced industrial democracies: Heterogeneity, transformation, and institutionalization. *Annual Review of Environment and Resources*, *40*(1), 337–361. https://doi.org/10.1146/annurev-environ-102014-021327

Greenberg, A., & Newman, L. H. (2020, May 31). How to protest safely in the age of surveillance. *Wired*. https://www.wired.com/story/how-to-protest-safely-surveillance-digital-privacy/

Grimm, J., & Harders, C. (2018). Unpacking the effects of repression: The evolution of Islamist repertoires of contention in Egypt after the fall of President Morsi. *Social Movement Studies*, *17* (1), 1–18. https://doi.org/10.1080/14742837.2017.1344547

Guariglia, M. (2020, November 5). *How to identify visible (and invisible) surveillance at protests*. Electronic Frontier Foundation. https://www.eff.org/deeplinks/2020/06/how-identify-visible-and-invisible-surveillance-protests

Guariglia, M., & Maass, D. (2021, February 16). *LAPD requested ring footage of Black Lives Matter protests*. Electronic Frontier Foundation. https://www.eff.org/deeplinks/2021/02/lapd-requested-ring-footage-black-lives-matter-protests

Jenkins, J. C., & Eckert, C. M. (1986). Channeling Black insurgency: Elite patronage and professional social movement organizations in the development of the Black movement. *American Sociological Review*, *51*(6), 812–829. https://doi.org/10.2307/2095369

Kalyvas, S. N. (2006). *The logic of violence in civil war*. Cambridge University Press.

Kanno-Youngs, Z. (2020, June 19). U.S. watched George Floyd protests in 15 cities using aerial surveillance. *The New York Times*. https://www.nytimes.com/2020/06/19/us/politics/george-floyd-protests-surveillance.html

Kazansky, B. (2021). 'It depends on your threat model': The anticipatory dimensions of resistance to data-driven surveillance. *Big Data & Society*, *8*(1), 205395172098555. https://doi.org/10.1177/2053951720985557

Klandermans, B. (1997). *The social psychology of protest*. Blackwell.

Kow, Y. M., Nardi, B., & Cheng, W. K. (2020). Be water: Technologies in the leaderless Anti-ELAB movement in Hong Kong. *CHI'20: Proceedings of the 2020 CHI Conference on Human Factors in Computing Systems*, 1–12. https://doi.org/10.1145/3313831.3376634

Lau, C., & Lum, A. (2019, November 8). Telegram social media channel used by protesters for doxxing Hong Kong police suspended after leaking thousands of pictures and videos of officers and their families. *South China Morning Post*. https://www.scmp.com/news/hong-kong/politics/article/3037001/telegram-social-media-channel-used-protesters-doxxing-hong

Lee, F. L. F., Liang, H., Cheng, E. W., Tang, G., & Yuen, S. (2021). Affordances, movement dynamics, and a centralized digital communication platform in a networked movement. *Information, Communication & Society*. https://doi.org/10.1080/1369118X.2021.1877772

Lee, F. L. F., Tang, G., Yuen, S., & Cheng, E. W. (2019). *Onsite survey findings in Hong Kong's anti-extradition bill protests*. Centre for Communication and Public Opinion Survey, The Chinese University of Hong Kong. https://sites.google.com/view/antielabsurvey-eng

Lee, F. L. F., Yuen, S., Tang, G., & Cheng, E. W. (2019). Hong Kong's summer of uprising: From anti-extradition to anti-authoritarian protests. *The China Review*, *19*(4), 1–32.

Liang, H., & Lee, F. L. F. (2021). Opinion leadership in a leaderless movement: Discussion of the anti-extradition bill movement in the 'LIHKG' web forum. *Social Movement Studies*, 1–19. https://doi.org/10.1080/14742837.2021.1989294

Lichbach, M. I. (1987). Deterrence or escalation? The puzzle of aggregate studies of repression and dissent. *The Journal of Conflict Resolution*, *31*(2), 266–297.

Liu, C. (2019). Multiple social credit systems in China. *Economic Sociology: The European Electronic Newsletter*, *21*(1), 22–32.

Li, Y-T. (2021). Accounting for "the social" in contact tracing applications: The paradox between public health governance and mistrust of government's data use. Big Data & Society. https://doi.org/10.1177/20539517211054277

Mahtani, S. (2019 , June 15). Masks, cash and apps: How Hong Kong's protesters find ways to outwit the surveillance state. *The Washington Post*. https://www.washingtonpost.com/world/asia_pacific/masks-cash-and-apps-how-hong-kongs-protesters-find-ways-to-outwit-the-surveillance-state/2019/06/15/8229169c-8ea0-11e9-b6f4-033356502dce_story.html

Mahtani, S. (2021, January 12). First came political crimes. Now, a digital crackdown descends on Hong Kong. *The Washington Post*. https://www.washingtonpost.com/world/asia_pacific/hong-kong-national-security-law-internet/2021/01/12/01738064-53b6-11eb-acc5-92d2819a1ccb_story.html

Marthews, A., & Tucker, C. (2017, February 17). Government surveillance and internet search behavior. Available at SSRN: https://ssrn.com/abstract=2412564

McAdam, D. (1990). *Freedom summer*. Oxford University Press.

McAdam, D., Tarrow, S., & Tilly, C. (2001). *The dynamics of contention*. Cambridge University Press.

McCarthy, J. D., & Zald, M. N. (1977). Resource mobilization and social movements: A partial theory. *The American Journal of Sociology*, *82*(6), 1212–1241.

Meyer, D. S., & Minkoff, D. C. (2004). Conceptualizing political opportunity. *Social Forces*, *82*(4), 1457–1492. https://doi.org/10.1353/sof.2004.0082

Milan, S. (2018). Political agency, digital traces, and bottom-up data practices. *International Journal of Communication*, *12*, 507–527.

Milan, S., & van der Velden, L. (2016). The alternative epistemologies of data activism. *Digital Culture & Society*, *2*(2), 57–74. https://doi.org/10.14361/dcs-2016-0205

Ng, K. C. (2020, January 8). Hong Kong police seized more than 3,700 mobile phones from protesters in space of five months and had devices broken into to read contents, security chief reveals. *South China Morning Post*. https://www.scmp.com/news/hong-kong/law-and-crime/article/3045263/hong-kong-police-seized-more-3700-mobile-phones

Opp, K.-D., & Roehl, W. (1990). Repression, micromobilization, and political protest. *Social Forces*, *69*(2), 521–547. https://doi.org/10.2307/2579672

Pang, J. (2019, June 21). Fate of 24 arrested Hong Kong protesters hangs in balance as anger turns on police. *Reuters*. https://www.reuters.com/article/uk-hongkong-extradition-protesters-idUKKCN1TM0Z8

Poupin, P. (2021). Social media and state repression: The case of VKontakte and the anti-garbage protest in Shies, in Far Northern Russia. *First Monday*, *26*(5). https://doi.org/10.5210/fm.v26i5.11711

Ray, L. O. (2021, September 3). Official emails show that LAPD worked with a controversial social media surveillance company during George Floyd protests. *LA Taco*. https://www.lataco.com/lapd-social-media-surveillance-protest/

Reny, T. T., & Newman, B. J. (2021). The opinion-mobilizing effect of social protest against police violence: Evidence from the 2020 George Floyd protests. *The American Political Science Review*, *115*(4), 1499–1507. https://doi.org/10.1017/S0003055421000460

Ringrose, K., & Ramjiee, D. (2020). Watch where you walk: Law enforcement surveillance and protester privacy. *California Law Review*, *11*, 349–366. https://doi.org/10.15779/Z38G44HR3X

Ruppert, E., Isin, E., & Bigo, D. (2017). Data politics. *Big Data & Society*, *4*(2), 205395171771774. https://doi.org/10.1177/2053951717717749

Schoolov, K. (2020, June 18). As protests over the killing of George Floyd continue, here's how police use powerful surveillance tech to track them. *CNBC*. https://www.cnbc.com/2020/06/18/heres-how-police-use-powerful-surveillance-tech-to-track-protestors.html

Sham Wing Kan v Commissioner of Police, HKCA 186. (2020). https://legalref.judiciary.hk/lrs/common/ju/ju_frame.jsp?DIS=127304&currpage=T

Siu, J., & Leung, C. (2020, August 28). Hong Kong protests: Alleged telegram channel administrator charged with inciting arson against police facilities, denied bail. *South China Morning Post*. https://www.scmp.com/news/hong-kong/law-and-crime/article/3099336/hong-kong-protests-alleged-telegram-channel

Smart Lampposts Technical Advisory Ad Hoc Committee. (2020). *Report of the technical advisory ad hoc committee on multi-functional smart lampposts*. https://www.ogcio.gov.hk/en/our_work/strategies/initiatives/smart_lampposts/committee/doc/CommitteeReport_en.pdf

Smith, T. (2019). In Hong Kong, protesters fight to stay anonymous. *The Verge*. https://www.theverge.com/2019/10/22/20926585/hong-kong-china-protest-mask-umbrella-anonymous-surveillance

Tarrow, S. (2011). *Power in movement: Social movements, collective action and politics* (2nd ed.). Cambridge University Press.

Taylor, V. (1989). Social movement continuity: The women's movement in abeyance. *American Sociological Review, 54*(5), 761–775. https://doi.org/10.2307/2117752

Tilly, C. (1978). *From mobilization to revolution*. Addison-Wesley.

Tilly, C., & Tarrow, S. (2015). *Contentious politics* (2nd ed.). Oxford University Press.

Ting, T.-Y. (2020). From 'be water' to 'be fire': Nascent smart mob and networked protests in Hong Kong. *Social Movement Studies, 19*(3), 362–368. https://doi.org/10.1080/14742837.2020.1727736

Tufekci, Z. (2017). *Twitter and tear gas: The power and fragility of networked protest*. Yale University Press.

Van Aelst, P., & Walgrave, S. (2004). New media, new movements? The role of the internet in shaping the 'anti-globalization' movement. In W. Van De Donk, B. D. Loader, P. G. Nixon, & D. Rucht (Eds.), *Cyberprotest: New media, citizens and social movement* (pp. 87–108). Routledge.

van Dijck, J. (2014). Datafication, dataism and dataveillance: Big data between scientific paradigm and ideology. *Surveillance & Society, 12*(2), 197–208. https://doi.org/10.24908/ss.v12i2.4776

Wade, K., Brubaker, J. R., & Fiesler, C. (2021). Protest privacy recommendations: An analysis of digital surveillance circumvention advice during Black Lives Matter protests. *CHI EA '21: Extended Abstracts of the 2021 CHI Conference on Human Factors in Computing Systems*. https://doi.org/10.1145/3411763.3451749

Whittaker, Z. (2021, February 7). Minneapolis police tapped Google to identify George Floyd protesters. *TechCrunch*. https://techcrunch.com/2021/02/06/minneapolis-protests-geofence-warrant/

Wong, B. (2021, April 20). Hong Kong protests: woman in doxxing case jailed for three years, becomes first person in two decades sentenced using colonial-era sedition law. *South China Morning Post*. https://www.scmp.com/news/hong-kong/law-and-crime/article/3130291/hong-kong-protests-woman-doxxing-case-jailed-three

Yuen, S., & Cheng, E. W. (2017). Neither repression nor concession? A regime's attribution against mass protests. *Political Studies, 65*(3), 611–630. https://doi.org/10.1177/0032321716674024

# Achieving organizationality through authorship affordances: a communicative episode of Telegram polling from 2019 Hong Kong

Marilyn Poon

**ABSTRACT**

In 2019, Hong Kong activists mobilized against the proposed Extradition Law Amendment Bill (Anti-ELAB) and relied on a mobile chat-app's polling feature to make collective decisions. To understand the chat-app's role in the movement's 'backstage' dynamics — i.e., building coalitions, making decisions, and coordinating action away from public view — I analyzed a series of Telegram polls during the Anti-ELAB movement's airport sit-in/takeover. To generate insights into social movement studies, I combined aspects of a vibrant stream of research known as Communication as Constitutive of Organizing (CCO) with a nuanced use of the *affordances* concept. To identify the backstage dynamics of the movement, I applied criteria from the organizationality concept to assess the extent to which the movement communicatively exhibited organizational characteristics. Therefore, I addressed how chat-app polling facilitated collective decision-making and cultivated the movement's social address and collective identity. Recognizing chat-app polling as pivotal in transforming voiced concerns to legitimized authority, I identified three authorship affordances that fostered authority practices: anonymity, hypertemporality, and affectivity. This study makes three contributions: (1) it extends the understanding of backstage organizational dynamics of social movements by adding two additional criteria of collective decision-making and collective actorhood to the extant criteria of collective identity, (2) it refines the affordances concept to explore how chat-apps influence data-related practices in contentious politics, and (3) it provides a novel method of analysis to isolate and investigate the processual dynamics of authority practices.

## Introduction

In the paradigm of datafication, networked activists are expanding their repertoire of communicative actions by appropriating and experimenting with smartphone applications (Treré, 2012). Activists' new forms of engagement with data has become a key area of social movement research, yet the bulk of this research has been concerned with the 'frontstage' (Goffman, 1959) — i.e., the visible aspects of social media platforms — including activists' and everyday platform users' sharing personalized content across platforms. The unprecedented opportunity to express themselves and find self-validation is commonly referred to as 'connective action' (Bennett & Segerberg, 2012, pp. 752–3). The research focus on these frontstage activities, however, overlooks the crucial 'backstage' mobilizing and organizing dynamics that shape movements' trajectories — i.e., how activists build coalitions, make decisions, and coordinate action outside of the public view (Treré, 2015).

Chat-app use is changing how communicative practices within contentious politics unfold. Recent empirical studies on the effects of chat-apps have shown that they can disseminate information (Kow et al., 2020); cultivate misinformation (Rogers, 2020); enable less visible, emotional, and even banal interaction that reinforces internal solidarity (Treré, 2020); and — because they facilitate frequent, almost-instant mobile interactions — constitute a new type of political activism (Milan & Barbosa, 2020). These studies have provided insights into how the practices activists engage in on chat-apps are different from their practices on publicly visible platforms (i.e., Facebook and Twitter/X). What they have not addressed, though, are the organizational dynamics of backstage activism, which remains undertheorized, specifically for democratic decision-making and collective identity formation (Gerbaudo & Treré, 2015). Although chat-apps are not always driven by the same algorithmic influence as social media is, their affordances — i.e., the actions they enable by design — are different from frontstage platforms, and these affordances significantly and discernably influence organizing and mobilizing interactions. In response to Milan's (2015a) call to avoid 'black boxing' technologies' agencies, this study seeks to understand their agential role in the organizational processes of contemporary social movements. My research question is therefore, how do chat-apps' communicative affordances help a social movement achieve the organizationality it needs to take collective action?

The Anti-Extradition Law Amendment Bill Movement (Anti-ELAB) is an appropriate context for examining how data-related practices help shape the evolving backstage organizing dynamics of contemporary social movements. Hong Kong has been undergoing an identity crisis since its transition from British colonial rule to China in 1997. As stipulated in the transition agreement, the promise of 'one country, two systems' will last until 2047, yet Beijing has already begun exerting authoritarian rule over the region. A notable example is the 2014 requirement that Beijing selectively pre-screen candidates for Hong Kong's chief executive. This action sparked the Umbrella Movement, encompassing groups of activists whose shared aim was to achieve more-transparent elections (Lee & Chan, 2016). Anti-ELAB began in response to a proposed extradition bill that would have allowed criminal suspects to be transferred to mainland China. The bill was perceived as further eroding Hong Kong's judicial independence. In 2019, spontaneous acts of contention took place in strategic locations in all districts of the dense city. Media coverage focused much attention on activists' use of chat-app polls — polls conducted on mobile devices using a chat application — in the dense yet heavily policed and surveilled city (Vincent, 2019). Created and shared within a chat, chat-app polls solicit users' opinions, preferences, or feedback on a specific topic or question. They are thus a promising avenue for investigating how activists strategically use data to transform collective grievance into collective action. This empirical study focuses on a series of polls carried out on the chat-app Telegram during the airport sit-in/takeover between

12 and 14 August 2019, a highly contentious period of the Anti-ELAB movement. The city incurred an approximate loss of US$76 million due to flight cancellations over these few days (Morrison, 2019).

This study contributes to social movement studies by drawing from a vibrant stream of organizational communication research known as the Communicative Constitution of Organizations/Organizing (CCO). CCO scholarship is united by the idea that organizations happen *in* and *through* communication (Cooren et al., 2011). From this perspective, (an) organization isn't solely a physical entity with a fixed location; it is also a dynamic process. Given this premise, understanding the communicative constitution of social movements — i.e., how they accomplish some degree of organization in and through communication — will provide valuable knowledge for social movement studies. The 'organizationality' concept, rooted in CCO (Dobusch & Schoeneborn, 2015), applies a minimalist understanding of what constitutes an organization. Through the analysis of four Telegram polls from Anti-ELAB's airport sit-in/takeover, I apply this concept to bring forth three organizational characteristics of the communicative episode: '(1) interconnected instances of decision-making, (2) actorhood, and (3) identity' (Dobusch & Schoeneborn, 2015, p. 1008). Recognizing chat-app polling as pivotal in transforming voiced concerns to legitimized authority, I identify Telegram's authorship affordances — the possibilities for communicative actions generated from actors' engagement with a technology — to relationally account for authority practices.

This study makes three contributions to social movement studies. First, it extends the current understanding of backstage organizational dynamics of social movements by adding two additional criteria of collective decision-making and collective actorhood to the extant criteria of collective identity. Second, it refines the affordances concept to better understand how chat-apps influence data-related practices — including the making of collective decisions — in contentious politics. Third, this study provides a novel analysis method to social movement studies that can be applied to isolate and investigate the processual dynamics of authority practices.

This paper is divided into four sections. The first section delineates the organizationality and affordances concepts. The second section provides the case selection and chat-app context. Section three details the study's empirical data and method of analysis. The fourth section analyzes the screenshots of Telegram polls of Anti-ELAB activists during the airport sit-in/takeover and reveals the three distinct authorship affordances for authority practices that emerged from the data: anonymity, hyper-temporality, and affectivity. Finally, to conclude, I identify my contributions and several implications of activists' engagement with data through chat-apps.

## Literature review

### *CCO & organizationality*

For CCO scholars, organization comes into being and sustains itself through a network of interconnected communication occurrences (Schoeneborn et al., 2019). This idea challenges the commonly held assumption that organizations are stable entities and opens a new way to study temporally and spatially contingent organizational phenomena. In this understanding, organizations are 'ongoing and precarious accomplishments realized, experienced, and identified primarily — if not exclusively — *in* communication processes' (Cooren et al., 2011, p. 1150, original emphasis). Hence, CCO theorizing can provides new insights on how social movements come into being and sustain through communicative practices (Schoeneborn et al., 2019).

By operationalizing the 'organizationality' concept, the study seeks to provide a communication-centered understanding of what it means when social movements organize. Dobusch and Schoeneborn (2015) developed the concept to investigate how loose social collectives can achieve varying degrees of organization based on three criteria: interconnected instances of decision-making, collective actorhood, and a collective sense of identity. The first criterion of *interconnected* decisions-making is one organization studies scholars consider 'the most fundamental aspect of organization' (Ahrne & Brunsson, 2011, p.85). Decisions are necessary to mobilize and organize social movements, and they also distinguish social movements from organizations because — in contrast to formal organizations — social movements need to make decisions about where, when, and how to meet (Haug, 2013). The aim of much of the organizing work in movements is about raising movement participants' consciousness and instilling the belief that they have a right to a voice. Whereas organizations do not need to continually make decisions about infrastructure, decision-making processes are integral to movement organizing, 'which works through developmental politics, in which the immediate objective may matter less than bringing people to see themselves as having the right and the capacity to have a say in the community or polity' (Gordon, 2002, pp. 104–5). . In chat-app polling, that voice is made up of separate framed decisions that are linked and aggregated, and that, collectively, speak for the movement. Yet implicit in this decision-making and voice of the movement is the authority necessary to act on those decisions.

The second criterion for organizationality is collective actorhood (Dobusch & Schoeneborn, 2015), which usually involves external actors attributing action by a specific social collective to a social address. This attribution process involves external actors appropriating — for themselves or others — the actions of something or someone else (Bencherki & Cooren, 2011). In technology-enabled movements, whose fluid forms of organizing lack the crucial boundary condition of membership common in formal organizations, a movement's ongoing communicative actions, such as user-generated platform posts, make it possible for external actors to identify the social entity as such. In other words, a movement can claim its own identity through repeated performances establishes this identity for external actors, who can then engage with the movement because it has achieved some degree of collective actorhood and attribute actions to it (Dobusch & Schoeneborn, 2015). Activists' engagement in 'contentious branding', built up through repeated performances of identity claims (Beraldo, 2020), gives way to a movement's organizationality. Importantly, though, while these communicative actions depend on the mechanisms embedded in the apps employed, they contain and realize movement relevant symbolic meaning.

The third criterion of organizationality is collective identity, a shared sense of belonging based on individuals' 'cognitive, moral, and emotional connection[s] with [the] broader community, category, practice, or institution' (Polletta & Jasper, 2001). Social movement scholars who used the resource mobilization and political process framework were unable to explain the emotional, socio-psychological and cultural factors influencing movement dynamics (Flesher Fominaya, 2010). The notion of collective identity helps these scholars overcome some of the inadequacies in their accounts because it offers an alternative explanation to material incentives, capturing the 'pleasures and obligations that actually persuade people to mobilize' (Polletta & Jasper, 2001). It further offers an alternative explanation to the making of instrumentally rational decisions (ibid.). Melucci (1996) suggests that digital technologies not only provide a means to achieve collective identity, but a direct pathway to change the symbolic and cultural framework through new forms of communicative practices and identity claims. This form of normative social control is accomplished 'through self-regulation and through the increasing penetration of standards of instrumental rationality into people's biological and emotional lives' (Polletta & Jasper, 2001, p. 286). To Milan (2015a), whose work builds on Melucci's, activists use data-based identity resources to 'select and emphasize meanings by their peers' to demarcate a boundary of 'Us against Them' (p. 895). Because of their enduring and spillover effects, collective identities have implications for societal dynamics, making it important to understand how apps aggregate individuals' meaning-making processes and shape the interactional dynamics of social movements.

From the CCO perspective, collective identity is achieved via the speech acts of heterogenous beings, specifically, through acts of authorship (Taylor & Cooren, 1997). In organizations, attribution traditionally happens when a group's representatives speak on its behalf (Bencherki & Cooren, 2011). In digitally networked protests, however, membership and leadership are often undefined, few — if-any — people expect representatives to speak for the protests. Instead, attributing, validating, and affirming the identity claims of a protest is done 'by grounding [those claims] in materiality,' including 'actual physical objects (artifacts), sites (such as well-established channels of communication), or bodies (i.e., identifiable human individuals)' (Dobusch & Schoeneborn, 2015, p. 1014). Authority is not held by an individual ex-ante but acquired over time and through interactions (Benoit-Barné & Cooren, 2009). By recognizing the agential character of texts, objects, and technologies, we may begin to understand what discursive implication these non-human actors (i.e., chat-apps) have on movement action. Taken together, the organizationality concept makes it possible to examine the constitutive elements of activists' communicative practices to shed light on the backstage dynamics of Anti-ELAB.

### From technological affordances to authorship affordances

The poll data from the key communicative episode of the airport sit-in/takeover were instrumental in mobilizing Anti-ELAB's collective action, an instance of 'data-as-repertoire' in which action, *enabled* by data, transforms the methods and circumstances of activism (Beraldo & Milan, 2019, p. 6). 'Data-as-repertoire' is a concept of data-related contention on one end of the 'contentious politics of data' spectrum (ibid.). From a data-as-repertoire perspective, activists use data as one of several tools in their repertoire of contention. On the opposite end is 'data-as-stakes', in which data becomes an intervention object that collective actors' use when reasoning and developing objectives (ibid.). The focus of this study is on data-as-repertoire, with the aim of revealing the communicative dynamics of Telegram polls in a specific event (the Hong Kong International Airport sit-in/takeover). Before we can understand the critical role of data in helping contemporary movements organize, we first need a nuanced conceptualization of the technologies that generate such data.

One way to conceptualize technology comes from ecological psychologist Gibson (1986), who developed the affordances concept to describe the perceived utility of objects by living beings. The concept has since been adopted, adapted, and extended by various disciplines, including media studies. When applied to digital technologies, affordances describe how their features interact with users' intentions to become the range of action possibilities. A nuanced concept of affordances helps to better understand how digital technologies both enables and constrains activists' capacity to construct meaning through collective experiences, and allows for the consideration of communicative practices within these mediation infrastructures.

Researchers in organization studies have applied the concept of affordances to the study of specific digital environments. Treem and Leonardi (2012), for instance, identified four social media affordances that shape users' behavior and lead to new organizational practices: *visibility* — the opportunity to make 'behaviors, knowledge, preferences, and communication network connections [...] visible to others'; *persistence* — the opportunity to make communication reviewable or recordable and accessible at a later time; *editability* — the opportunity to 'revise, reshape, and coordinate content'; and *association* — the opportunity to 'established connections' between actor(s) and content. Similarly, Zammuto et al. (2007) used the concept of *affordances for organizing* to simultaneously analyze the features of digital technologies and organization systems, and found that context links the two: '[O]ne cannot talk about a complex technology without reference to the social setting' (pp. 752–3), because a digital technology's specific functions were related to individuals' social competencies and dispositions towards organizing. Scholars have continued to apply the concept of affordances by showing how they are inextricably linked to practice. Faraj and Azad (2012), for instance, have shown that the affordances concept makes it possible for materiality to be taken seriously when studying social interactions, and they define 'technology affordances' as 'action possibilities and opportunities that emerge from actors engaging with a focal technology' (p. 238). Emphasizing a practice-based view, Fayard and Weeks (2014) extended the focus of Faraj and Azad, from the 'affordances *of* technology to the affordances *for* practice' (p. 236; original emphasis). This conception combines the possibilities of technology with how its materiality is actualized in practice, to challenges the taken-for-granted distinctions of the sociality and materiality. Incorporating a CCO perspective to take the agential character of technology seriously, this study applies a nuanced concept of authorship affordances for authority practice to how a chat-app can act on the activists' decision-making data and influence authority practices.

Focusing on Telegram's authorship affordances helps identify which communicative actions Anti-ELAB protesters had available to them in the backstage organizational dynamics. However, it also helps reveal the effects of these actions and the power mechanisms inherent in the app's polling feature. This focus on authorship affordances thus responds to Cammaerts (2012) call to recognize the discursive dynamics in movement action, since the various technologies that protesters use increasingly sustain movements. He proposes viewing the interrelated dynamics of activism as embedded within a 'mediation opportunity structure' to accounts for the relationships between various media actors' representations of activists, activists' self-mediation processes, and technologies. Cammaerts' (2012) concept delineates three non-exclusive protest action logics that empower and constrain activists: (1) logic of numbers in terms of presence in demonstrations, (2) logic of damage in terms of property destruction or disruption to societal order, and (3) logic of bearing witness in terms of public performance that decries injustice and civic disobedience. While the above concepts have significantly advanced the understanding of movement dynamics in the digital age, they cannot clearly explicate the organizational dynamics of social movements.

## Case selection & Telegram

The press reported that Anti-ELAB activists extensively used Telegram and data-based tactics ('data-as-repertoire'; Beraldo & Milan, 2019) to escalate protest activities all over Hong Kong, a semi-autonomous region of China. This case examines one such tactic — making collective decisions using Telegram polls. The movement resembled a network and has been characterized as a smart mob because activists used new mobile apps and 'like water' tactics — adaptability, flexibility, and fluidity (Ting, 2020). While there is truth to this characterization, it does not reveal the mechanisms and processes activists needed to mobilize and organize. The aim of this study is, therefore, to reveal these mechanisms and processes by showing how the authorship affordances of Telegram made it possible for Anti-ELAB activists to establish a set of authority practices and achieve organizationality in a pivotal communicative episode.

Telegram's encryption features made the app attractive to Anti-ELAB activists. Downloads of the app surged in July 2019 in Hong Kong, indicating just how widespread its use was during a highly contentious point in the movement (Banjo et al., 2019). Anti-ELAB activists used more than 100 groups and channels — some with over 10,000 subscribers — to post news of upcoming protests, to provide tips on how to douse tear gas and dox suspected undercover police, and even to share access codes that allowed protesters to enter and hide in buildings (Lee, 2020). The use of polls to make collective strategic decisions was a new tool in the repertoire of movement's communication, a practice that at least seeks to align with the notion that everyone gets a vote and maintains some form of participatory democracy.

While groups on chat-apps — like WhatsApp — are usually formed amongst a group of users with established social ties, Telegram channels function by establishing ties among a network of users who are not necessarily previously connected. The right to post messages or polls is limited to channel owners or admins, a privilege 'affording a form of broadcasting' (Kow et al., 2020, p. 4). It is a 'hybrid system' that seeks to protect the privacy of its users while allowing them to communicate with a large group (Rogers, 2020). Any Telegram user can start a public or private channel with no limit to the number of subscribers. Any user can join a public channel by searching for it by name, but joining a private channel requires the approval of an administrator (admin), who must either approve requests individually or send users a link to join. A channel's owner decides who gets to be an admin and can give admins additional rights, such as adding new admins.

## Data and research method

This research began in early 2020, before the opaque national security law was announced in Hong Kong. Many Telegram channels in use during the protests were already defunct, and my attempts to access a more substantial amount of data through the channels that remained were unsuccessful. Still, I continued this research by using data published in a news story on Bloomberg, a reliable international news agency (Banjo et al., 2019). The dataset comprises four Telegram screenshots of activists' polled decisions. The screenshots visually represent what activists saw on their mobile devices during the events. I closely analyzed these naturally occurring interactions to understand the achievement of authority in chat app polling communication (Benoit-Barné & Cooren, 2009). These four polls are from a pivotal episode of Anti-ELAB, when action shifted from peaceful sit-in to heated violence at the airport, changing the character of the movement. I corroborated Bloomberg's accounts of the events through close readings of various news reports and found no issues with its authenticity.

From this select sample, I was able to explore Telegram's channel feature and polling mechanism. Analyzing these aspects helped me carry out the study's objective of delving into the fundamental process of meaning-making within the communicative constitution of organization(ality). Studies such as this one focusing on small data are valuable because they can be used to make 'targeted' investigations that reveal the unique perspectives in 'individual, nuanced, and contextual narratives' (Kitchin & Lauriault, 2015, p. 466). Anti-ELAB activists used other chat-apps in addition to Telegram during the protests to carry out polls, including the Reddit-like forum LIHKG, where users can vote content up or down. Yet I consciously limited the scope of this study to these four Telegram polls to answer my research question.

An important ethical consideration for researchers is that chat-app users expect that their exchanges are private. Furthermore, in the context of political activism, these users are also vulnerable people (Barbosa & Milan, 2019). By default, usernames are not visible on Telegram's user interface. Although identifying individuals by channel names alone is unlikely, I have masked the channel names on the screenshots to prevent users from being traced.

Using a CCO approach, this study focuses on the authorship affordances from activists' engagement with Telegram channels, those which constitute movement organizationality. To identify patterns of interaction in their communicative practices, I qualitatively analyze the content of a series of Anti-ELAB activists' polled decisions. The CCO understanding of how *matters of concern* gain authority makes it possible to isolate and analyze the conversations and texts that end up mattering more to subscribers in the chat-app polls from a key Anti-ELAB communicative episode (Cooren et al., 2015). Matters of concern are the fleeting moments in communicative episodes (ibid.), the moments that 'drive participants to defend or evaluate a position, account for or disalign with an action, or justify or oppose an objective' (Vásquez et al., 2018). In activism, participants can invoke many concerns, by shouting them out as a slogan, writing them onto posters, or posting them on digital platforms. In my analysis of how discursive resources are mobilized to enact resistive practices, the use of the matters of concern to matters of authority framework serves to isolate and examine the flow of action in which actors' specific concerns gain authority.

All conversation and text in Telegram channels, including the polls posted by channels' admins and the subscribers' voluntary responses to these questions, are *matters of concern*. Once a poll has been taken, Telegram algorithmically tallies individual choices, then almost instantly authors — and simultaneously completes the authorization of — a collective decision, whereby one specific concern becomes *a matter of authority*.

## Findings and discussion

I first present the four Telegram polls sequentially, along with my translation from Chinese to English, and then interpret these polls in the context of events surrounding the movement. Afterward, I describe the three authorship affordances which emerged from the data: *anonymity*, *hyper-temporality*, and *affectivity*. Using the concept of authorship affordances for authority practices from a communication as constitutive perspective reveals how chat-apps influence data-related practices — specifically, the making of collective decisions in contentious politics.

### *Poll 1 — 'agree with escalating airport action?'*

Around midnight on 11 August, a poll was created asking whether the action at the airport, where a peaceful sit-in had begun three days earlier, should be escalated (Lee, 2020). Just a day prior to this poll, a protester was shot in the eye by a beanbag round and sustained serious injury, which further angered activists. Their plans to disrupt airport operations can be seen as a response to the increased police violence and lack of police accountability for unjust actions in recent events.

The poll question, 'Agree with escalating airport action?', had two possible responses: 'agree' or 'oppose'. The prompt 'this instant!' following the question added a sense of urgency, and the emotional expression 'angry to the point of explosion' following the 'agree' option reflects activists' general sentiment. In this first poll, 79% of respondents voted in favor of escalating the airport actions. The message that follows the first poll reminded voters that the plan to enter the airport had its origins in an earlier idea to create disruptive activities (see figure 1).

### *Poll 2 — 'whether or not to allow that Chinese police officer to leave?'*

On 13 August, with protesters still at the airport, hundreds of flights were cancelled and news outlets reported that the protesters at the airport had detained and assaulted an undercover Chinese police officer (Banjo et al., 2019). At the time of the poll, the police officer was injured and required medical attention (ibid.). The poll question begins with 'one sentence to conclude', a colloquial saying the admins used to indicate that it was time to end the conversation. The question 'Whether or not to allow that Chinese police to leave?' is direct and logical, and is strategically effective for bringing to a close the various matters of concerns that had been raised and for authorizing a collective decision.

The person held captive by the protesters is referred to as 'Kung On', a term commonly used for mainland Chinese police. Also notable is that of the four options, the two using slang terms received the most votes, 49% and 25% (see figure 2).

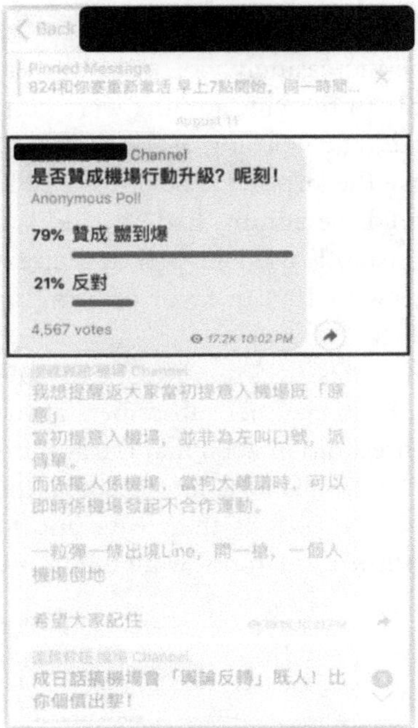


[Name obfuscated:] Chinese Idiom Channel

Agree with escalating airport actions? This instant!

Anonymous Poll

79% agree angry to the point of explosion

21% oppose

4,567 votes

17.2K 10:02PM

I want to remind everyone that the 'original idea' to enter the airport was not to call out slogans or pass out flyers.

It is to place people at the airport so that when the dog crosses the line, we can instantaneously stir up disrupting activity.

If there's one bullet or one person crossing the line, fire a gun, and one person will fall at the airport.

Hope everyone remembers.

**Figure 1.** Poll 1: 'Agree with escalating airport action?'.

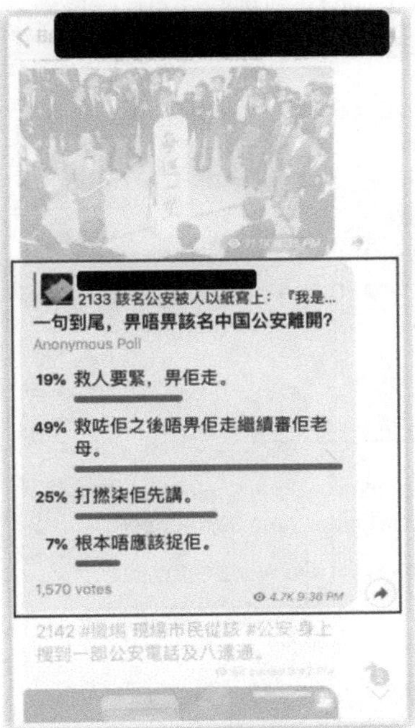


[Name obfuscated:] Irrelevant action verbs

[image of a mock detainment]

One sentence to conclude, whether or not to allow that Chinese police officer to leave?

Anonymous Poll

19% Saving life is important, let him leave.

49% Save him but do not let him leave after and continue to interrogate his mother [slang term].

25% Beat him [in slang] and then discuss.

7% From the beginning should not have captured him.

1,570 votes

4.7K 9:36 PM

#airport onsite citizen found in #Kung On_possession a Chinese police phone and subway card.

**Figure 2.** Poll 2: 'Whether or not to allow that Chinese police officer to leave?'.

### *Poll 3 — 'should there be calls to leave the airport?'*

Poll 3 was also conducted on 13 August, which became increasingly violent as the day went on. Recognizing that heightened police action was imminent, admins were concerned about activists' safety and thus called for this vote. The wording of the third poll question is clear and direct: 'Should there be calls to leave the airport?' The phrase '2000 votes election' following the question refers to a threshold the admins had set, and is a noticeably different discursive device from the 'this instant!' used in poll 1. Retreat (leaving the airport) was the choice of 81% who voted, with 19% in favor of bringing in more support (i.e., additional protesters) to the airport.

The text beneath poll 3: 'Referendum results, 3000 votes in 5 minutes, with 50% participation, and 80% support', reveals that there were 6000 channel subscribers at the time of the vote, and therefore that the minimum vote threshold of 2000 that the admins set represented about one-third of the channel's subscribers. This post summarizing the authorized results also urges everyone to not damage the airport as they leave and to avoid clashing with the police as they do so by ending with, ' … Do not want any more harm done to arms legs [comrades]!' (see figure 3).

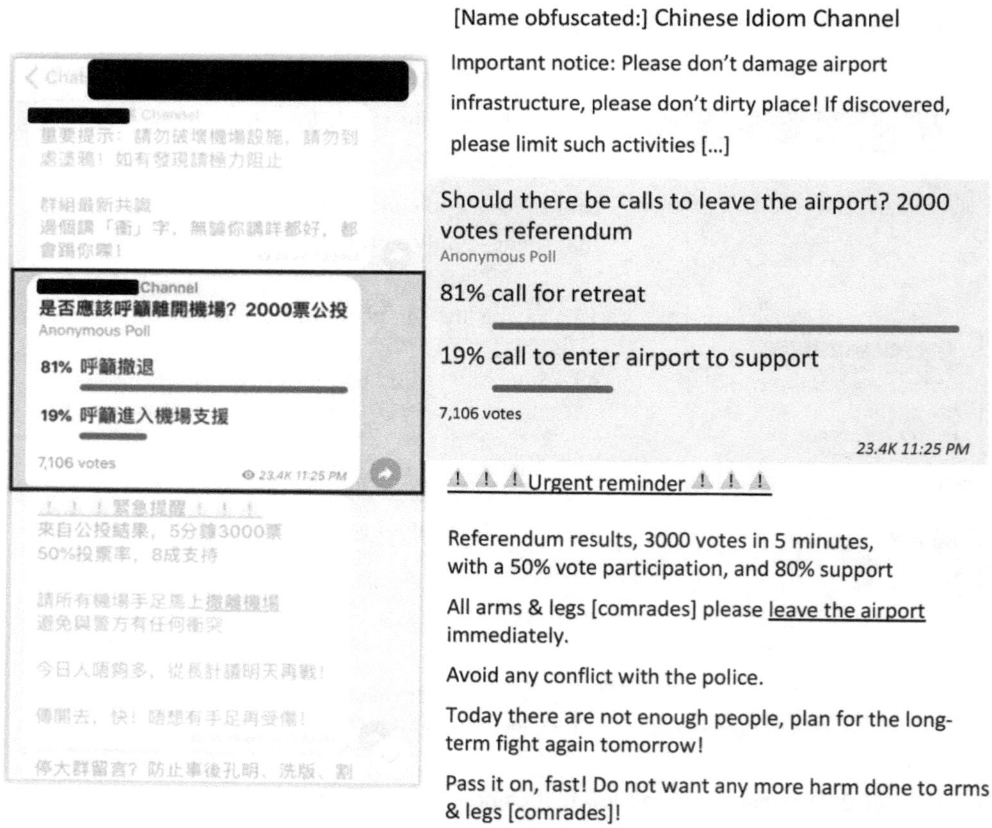


**Figure 3.** Poll 3: 'Should there be calls to leave the airport?'.

### *Poll 4 — 'Tomorrow apology day?'*

The violence that transpired throughout the day on 13 August brought negative attention from various media outlets. Paramedics eventually evacuated the detained Chinese police officer from the airport so he could receive medical attention. Poll 4, calling for protesters to decide whether to apologize, likely took place in the evening. Phrased simply, 'Tomorrow apology day?', it provided subscribers with two options: agree or disagree. The language in this poll is neutral, with no narrative devices of slang or emotional qualifiers. At the time the screenshot was captured, this poll had received 10,709 votes, with 74% in favor of apologizing and 26% opposed.

Of special note in poll 4 is the message that precedes it, urging admins to honor a period of election silence between 18:00 and 23:00 by not posting messages. The authoring admins added this message because they wanted the poll to remain visible to other users, since any new posts would push the poll further down the feed and might prevent some subscribers from seeing it. The message that follows, 'No matter if we do it or not, after the incidents others would demand us to explain, hence, everyone vote!', shows that the admins expected that activists would be held accountable for the violence at the airport (see figure 4).

As reviewed earlier, identities claims are grounded in materiality (Dobusch & Schoeneborn, 2015), therefore, to understanding how Anti-ELAB achieved organizationality, it is necessary to examine how decision- and claim-making practices are anchored to materiality — from the actual devices used by the activists to the digital infrastructure

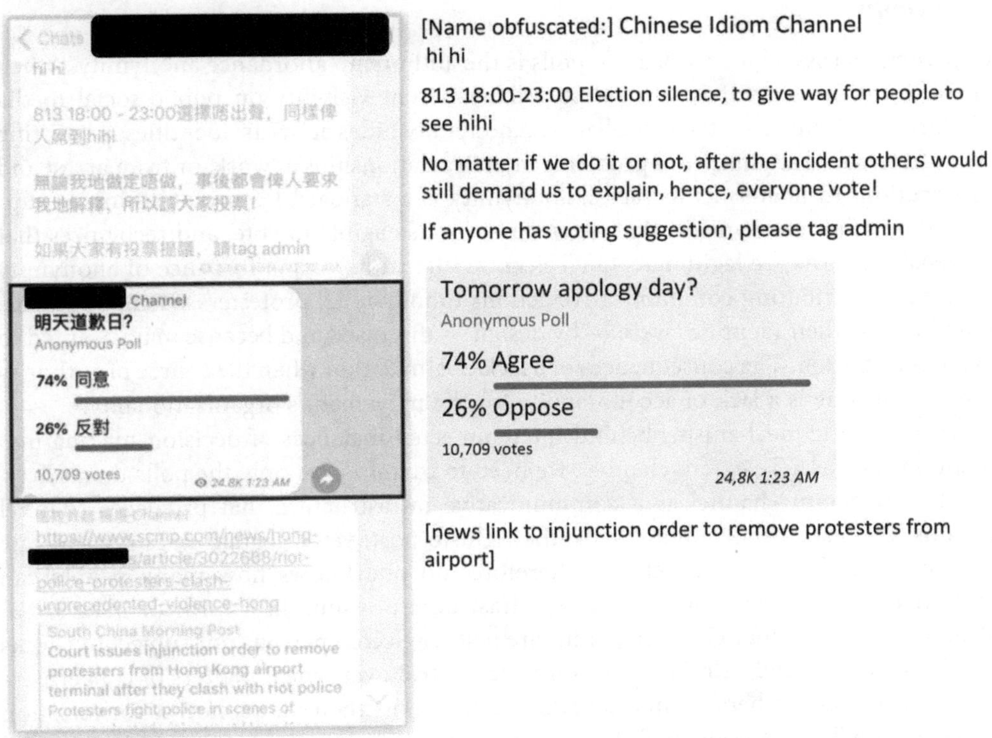


**Figure 4.** Poll 4: 'Tomorrow apology day?'.

**Table 1.** The occurrences of the three authorship affordances (anonymity, hyper-temporality, and affectivity) in the four Telegram polls.

| Authorship Affordances for Authority Practices | | | | |
|---|---|---|---|---|
| Poll | Textual devices applied in Telegram Polls by channel admins | Anonymity | Hyper-temporality | Affectivity |
| 1 | Term 狗 'dog', a dehumanizing term for Hong Kong police | x | | x |
| 1 | Colloquial saying 'angry to the point of explosion' | x | | x |
| 1 | Prompt: 'This instance!' | x | x | |
| 2 | Colloquial saying 'One last sentence' | x | x | |
| 2 | Term 共安 'Kung on' to refer to Chinese police officer | x | | x |
| 2 | Slang term 'continue to interrogate his mother' | x | | x |
| 3 | '2000 vote' threshold on 'referendum' | x | x | |
| 3 | Term 手足 'arms and legs' to refer to comrades | x | | x |
| 4 | Use of 'we/us' to refer to in-group, and 'people' to refer to the public | x | | x |
| 4 | Term 大家 'big' and 'family' to refer to everyone | x | | x |
| 4 | The specification of a five hour voting window | x | x | x |

to the individual bodies embodying intentions and emotions. Combining these searches with the attention to the chat-app revealed three authorship affordances for authority practices: anonymity, hyper-temporality, and affectivity (see table 1).

## *Anonymity*

Consistent across all four Telegram polls is the authorship affordance anonymity, which enabled anti-ELAB activists to sidestep the inherent visibility on public social media platforms and thus counter surveillance. Anonymity hides activists' identities, protecting them from social stigma, from problems in their relationships at work, or from arrest and prosecution. In liberal democracies, anonymity is a standard feature of voting, is supported by other mechanisms that determine who is eligible to vote, and recognizes that citizens' concerns are legitimate. On Telegram, the authorship affordance of anonymity meant that attributing communicative actions to individual protesters was not possible, both because their identities were — by design — disguised and because many wanted to avoid recognition. The consequences of a lack of attribution when the source of authority remains opaque is a lack of accountability for the movement's organizationality.

To reveal the mechanisms behind interconnected instances of decision-making in a medium such as a Telegram channel, we need to examine not only the polls themselves, but the Telegram channel as a communicative infrastructure that privileges channel admins. By design, only admins have the right to post, yet Telegram does not indicate how many admins a channel has; therefore, no one knows how many people can authorize polls or who they are. This infrastructure results in a form of inadvertent stealth leadership, since channel admins are neither elected persons nor subject to checks and balances. Overall, admins have complete control over which matters of concern are polled. Since subscribers do not have the right to post their views, how a conversation unfolds on a Telegram channel demands scrutiny. The infrastructure of Telegram and the interactions it produces contradicts with a core tenet in the *matters of concern to matters of authority* framework, which is that participants should be able to voice and negotiate matters of concern (Vásquez et al., 2018). In effect, subscribers' voices are sidelined in Telegram channels, a grossly imbalanced arrangement of authoring power. The consequence of this affordance to authority practices echoes what Hensby (2017) observed in the 2010–11 UK student protests, in which a core group of activists began using Facebook 'secret' groups — not searchable or visible to non-members — to organize. The hierarchy that emerges from certain activists' online authority practices runs counter to activists' initial commitment to, and expectation of, horizontality and openness (Hensby, 2016).

The anonymity of admins contrasts with the conventional social media affordance of visibility (Treem & Leonardi, 2012), which still affects Telegram polling, since the votes of individual subscribers are visible. At first glance, visibly displaying the number of votes aligns with the expectation that online interactions are transparent, adding legitimacy to chat-app polling. However, Telegram polls impose a logic of numbers, in which a mass presence of activists produces a spectacle that the media can report on (Cammaerts, 2012). In addition, the vote counts, the number of people who have seen the poll, and the number of people who have voted are displayed in real time, which can influence individual's voting choice. Since collective behavior affects rational decision-making in the form of a 'threshold,' which varies from individual to individual, making the popular choice visible increases the likelihood that individual decisions are influenced by a 'bandwagon effect' (Granovetter, 1978). Hence, Telegram's vote-counter mechanism may produce a mob effect that silences less popular opinions. As activists see the voting results in real-time, some may limit their choice to the top two options (in polls with three or more options) so that they can have a salient impact. Taken together, the authority practices facilitated by Telegram seem transparent, yet their very visibility can have unexpected and undemocratic implications. The politics of visibility and representational dynamics shape and determine the identity-formation process, and the degree of organizationality enabled through cloud protesting 'do[es] not impose the *degree of responsibility* towards fellow activists that real-life groups do' (Milan, 2015b, p. 7, emphasis added). Thus, the volume of data made visible to Anti-ELAB activists on Telegram imposes a logic of quantification that influences their perception and manipulates social dynamics. This influence and manipulation likely affected Poll 2, when activists decided to detain and beat the Chinese police officer.

Another troubling aspect of the authorship affordance of anonymity is that it is not genuinely anonymous: the data created can be traced back to the phone number to which the app is linked. Hong Kong authorities, for example, have requested user data from Telegram. Although the company has refused to comply, users have no guarantee that their data will always remain only with Telegram. The national security law enacted in Hong Kong in June 2020, combined with the shifting influence and impact of politics on tech enterprises, means that user data can be handed over to authorities in the future, and they can use it to identify and prosecute users. Many activists preempted this threat using pre-paid SIM cards to activate and use Telegram (Vincent, 2019).

### *Hyper-temporality*

The second authorship affordance of Telegram that the data revealed is hyper-temporality — an urgency and immediacy that influenced the protests. 'Hyper' refers to a heightened intensity of temporal logic that links and strengthens other logics: the temporal logic of Telegram that made real-time voting possible, for example, is linked to the logic of quantification that displayed instantaneous vote counts. The logic of numbers thus heightens the temporal logic, which caters to chat-app users' expectation of quick responses, even more so than with other apps. While the orientation of mobile apps tends to be urgent, chat-app have an even greater hyper-orientation.

The temporal logic of immediacy shapes how activists organize for collective action on Telegram, yet it does not *determine* activists' mode of interaction. In Poll 1, for instance, the channel owners and admins urged activists to vote 'this instant', thus exploiting the logic of urgency inherent in the app. In contrast, in Poll 4, they attempted to modify that logic by imposing impose 'election silence', thus providing an alternate logic in authoring.

With the statement 'One sentence to conclude' in Poll 2, the admins applied an alternate temporal logic to end the conversation. This colloquial saying served to signal a stop of input of various matters of concerns by admins, to denote a temporal transition to collective decision-making with the poll question ' ... whether or not to allow that Chinese police to leave?' While not contradicting the chat-apps' logic of urgency, this saying served as a discursive resource that reinforced the sense of collective identity as the authoring admins signaled their belonging-ness to the activists community while practicing authority.

In Poll 3, however, the admins sought to combine logics to influence how polls are conducted. First, the admins added their quantification logic to the Telegram logic of urgency with the prompt '2000 votes election' following the poll question 'Should there be calls to leave the airport?' The threshold admins set in this poll of having at least a third of channel subscribers vote — a threshold that was exceeded (over 50% of subscribers voted) in five minutes — added to the dominant temporal logic of the chat-app and led to a rapid response. In other words, the admin-determined vote threshold hyped the prevailing logic of urgency. The effects of these logics on the outcome of a collectively authored decision reveal the materiality of chat-app: in contributing to enacting authority practices, the chat-app shares agency. Because the collectively authorized decision was only possible because of these logics, the chat-app ultimately shared authorship in the decision, confirming that non-human agents can 'end up being more powerful and authoritative than others' (Benoit-Barné & Cooren, 2009).

In Poll 4, the admins more explicitly countered the hyper-temporality authorship affordance. The message preceding the poll — about whether to apologize tomorrow — asks admins not to post for five hours so that the poll will be more visible on the feed, and more subscribers can see it and vote. In countering the dominant urgent temporal logic of the app by extending polling time, admins gave subscribers more time to make a more-considered decision. The decision in this poll is partly a response to activists' regret for their decision in poll 2 to detain and physically harm the Chinese police officer. It is not clear whether the admins' five-hour polling window influenced the results, but that the admins felt compelled to resist the temporal logic with such a large window could reasonably be interpreted as being driven by a sense of responsibility and regret. If so, countering temporal logic and hyper-temporality in authorship affordances signify to subscribers the gravity of their decisions. It also shows that the admins have reservations about how polled decisions have influenced the trajectory of the movement.

### *Affectivity*

Affectivity, the third authorship affordance of Telegram revealed in the data, influenced decision-making and helped to give the movement its identity. In each of the four polls, the vote options that used emotional wording received the most votes. Affectivity thus transformed various matters of concern into matters of authority. Research has well-established that emotional talk can trigger embodied affection and influence decision-making. The common perception is that negative online content is more impactful (Gerbaudo et al., 2019), but in these four polls admins used a combination of discursive and affective resources — such as inculcating a sense of belonging and using an 'us vs. them' narrative — to formulate the polls and answers, and thus to frame the movement's beliefs and meaning, from which it formed its identity (Jasper, 1998; Polletta & Jasper, 2001; Polletta, 2006).

In Poll 1, the option to agree was followed by 'angry to the point of explosion.' Admins used this heated, affectual expression to foster a sense of belonging among activists. Interactions such as this one, which exhibit emotional investments, are essential to negotiating identity because they 'enable individuals to recognize themselves in each other' (Melucci, 1989, p. 35). When individuals' identities are at stake, affective language can unify and reinforce the internal solidarity of loose social collectives (Treré, 2020). In this case, it was not just that the language was affective, but that it was communicating a symbolic nuance to an 'in group' that would recognize its significance. 'angry to the point of explosion' is a Cantonese saying, in this poll written in traditional Chinese (as opposed to the simplified Chinese used in mainland China), which made it an identity resource for Anti-ELAB activists. Such narrative tools work because they are drawn from a mnemonic community's emotional and cognitive repertoire to shape its identity (Wertsch, 2012). The poll options using affectivity drew more votes because admins were speaking on behalf of Anti-ELAB activists and appealing to their identities.

In Poll 2, the term 'Kung On' was used as an 'us vs. them' narrative tool. Translated, it means 'public peace', a term used only for Chinese police officers from mainland China. Using this term distanced activists from the captive police officer by aligning him with Chinese authorities, marking him as the *other* to reinforce their Hong Kong identity as distinct from that of China. This sense of belonging and having a distinct identity was present throughout the protests, with admins referring to movement activists as comrades and using the Chinese characters 'arms' and 'legs' in Poll 3, which symbolically identified fellow activists as part of a body to foster a sense of belonging among members of the movement and to maintain solidarity. This term was continually re-authored in talk and text, performatively reinforcing Anti-ELAB's coherent actorhood. Similarly, admins referred to Hong Kong police officers as dogs in the first post and throughout the protests. This dehumanizing term had its counterpart in Beijing and police supporters' use of 'cockroach', perhaps an even more dehumanizing term to refer to Anti-ELAB activists. Using these terms also underscores the significant impact of framing and — importantly — who is doing the framing. Forming a collective identity in protests requires that shared personal frames resonate with the collective. Resonance is created through actors' interactions, and the authority of that resonance increases as texts are re-authored through various online text and talk channels. On Telegram channels, however, only admins were in the powerful position to frame, package, present, activate, and coproduce this authority since, by design, they control channel broadcasting rights, and thus controlled the infrastructure to determine the knowledge-construction processes (Milan & van der Velden, 2016).

The power embedded in the authority practices is seen in the affectivity that was present in the wording of two of the voting options in Poll 2, which most likely influenced the results since these two options received the most votes. With 49% of the votes, 'Save him but do not let him leave after and continue to interrogate his mother (Cantonese slang term)' became a matter of authority, even though it was contradictory. Carrying it out would have resulted in two inconsistent actions: letting him leave the airport to receive medical care (and thus be saved) and prohibiting him from leaving so that he could continue to be interrogated. Perhaps the admins expected that medics would enter the airport, treat him, and leave, and the activists could then continue to interrogate him. However, even if they expected this turn of events, the affectivity in the admins' language was clear. The option with the second-most votes, at 25%, also used inciteful, affective language: 'Beat him (in slang) and then discuss'. In comparison to these options, the one with the third-most votes (19%) is reasonable and logical: 'Saving life is important, let him leave', while the option with the least number of votes (7%) — 'From the beginning should not have captured him — denotes a sober regret of the current situation.

Poll 2 thus clearly marked a shift in activists' thinking and emotional state, as well as in the movement's actions and direction. Up until this communicative episode, the movement had, by and large, been non-violent. The rallies, general strikes, and police station blockades had been peaceful. Even when activists took over the Legislative Council in July, only property was damaged (Morrison, 2019). This violent shift in the options in Poll 2 May has been an emotional reaction prompted by the Yuen Long incident on 21 July, when masked men, thought to be triad/gang members, attacked protesters and citizens while the police looked the other way (Lee, 2020). The sequence of the polls also suggests that the internal narrative of the movement was one of emotional response followed by questioning and regret since, according to Banjo et al. (2019), the third poll about whether to leave the airport took place around the time the activists detained another person. 'The soul-searching began shortly afterward' (ibid.), as the fourth poll on whether to apologize the next day took place the same night.

Admins played on Anti-ELAB's established ethos of maintaining solidarity in poll 4 to drive the collective towards making the sensible decision to apologize. In the message preceding the poll question, the admins wrote that 'people' (i.e., the public) would demand an explanation from the collective for what had transpired at the airport. This message primed voters by establishing external expectations, promoting a sense of solidarity, and positionin the proposed apology as grave and critical. The term the admins used for 'people' is significant here since this neutral term does not make a distinction between 'us' vs. 'them.' In addition, the *everyone* in 'everyone vote!' was also meant to instill a sense of cohesion. Composed of the two characters 'big' and 'family,' this way of expressing *everyone* contrasts with another more-formal term in Chinese composed of the characters 'each' and 'position.' Thus, although the admins did not ostensibly use emotions in this poll, their choice of characters were instances of using affectivity to reprioritize solidarity, also from the public's perspective.

## Conclusion

As this study illustrates, the organizationality concept supplements the well-established notion of collective identity with collective actorhood and inter-connected instances of decision-making, extending the understanding of a movement's backstage organizational dynamics. The three authorship affordances this study identifies authorized collective action in chat-app polls, thus revealing how algorithmically structured interactions transformed Anti-ELAB organizationality. The chat-app Telegram thus 'interven[ed] in the meaning-making process by means of their algorithmic power' (Milan, 2015a, p. 888) and shaped Anti-ELAB's authority practices.

This study found three specific authorship affordances that fostered authority practices *in* and *through* communication: anonymity, hyper-temporality, and affection. The authorship affordances of digital apps in this evolutionary communicative process are important because they determine who can speak, how they speak, and what can be spoken, directly impacting the performance of identity claims. This study contributes to the existing literature by showing how a nuanced concept of affordances can be applied to study the communicative dynamics of organizations/organizing/organizationality. Moreover, this study has show how the matters of concern that the collective authorize have implications on its collective actorhood and internal solidarity, which binds its collective identity. Therefore, this study highlights the usefulness of the three criteria to the organizationality as complementary to the widely acknowledged notion of collective identity in social movement studies.

This study revealed that authoritative power is a negotiation between activists and technology. Unnamed and unknown admins were the only ones with the right to author a poll text, creating an imbalance of power between subscribers and admins. This lack of transparency is equivalent to a '*loss of accountability*' (Weiskopf, 2020). Although Telegram channel subscribers have the right to participate in polls, thus gaining a sense of belongingness, their voices are never made present. In addition, the quantifiability in Telegram, in the form of the tallies of votes and voting responses in the polling mechanism, had a determinative influence on how Anti-ELAB activists could and did respond, leaving them with a '*loss of response-ability*' (ibid.). Telegram polling transformed Anti-ELAB activists' grievances and emotions into action. However, it did so at a very human cost because it prevented activists from considering the impacts of their decisions. It was impossible for activists to tangibly understand the consequences of their actions, with regrettable results. As Weiskopf rightly stated, 'The ability to respond to the concrete other is a precondition for exercising or enacting moral responsibility' (ibid., p. 24). We should be concerned with how specific affordances shape our ability to response because a movement that seeks to achieve collective action should consider the morality of its practice. Achieving this accountability is not easy; in authoritarian regimes, identifying admins and users is not an option, but there may be other ways to overcome forms of stealth power. As a start, activists can acknowledge that the popular characterization of their movements as 'leaderless' is an affective attachment that masks inherent power dynamics (Poon & Kohlberger, 2022). Making an important political impact means making power *visible*, and therefore negotiable (Melucci, 1985).

A final consideration for activists based on this study's findings is whether their data practices contribute to a data governance regime. Activists should be aware that their data may endanger themselves and that they may be liable to suppression in the future. While most activists see data as one tool in a repertoire, they should be mindful that data can later become 'stakes' in contention (Beraldo & Milan, 2019). Many apps build in encryption and privacy mechanisms, and Telegram has a track record of advocating and supporting free speech. Yet, political activists should use untraceable SIM cards as a precaution since business policies, practices, and owners can and do change. A viable alternative is using open-source decision-making chat-apps rather than profit-driven ones.

While this study provides opportunities and implications for future research, it also has limitations. An important question for researchers is how datafication and other media technologies enable or constrain a movement's collective actions, which researchers can investigate by combining the concepts of authorship affordances with organizationality used in this study. A limitation of this study was using news and academic sources to closely analyze archived digital objects as representations of naturally occurring interactions. Researchers should also expand the scope of their data sources beyond these sources by investigating activists' actions on the ground.

## Acknowledgements

I am grateful for the opportunity to contribute to this special issue as an organization studies scholar. I would like to thank Yao-Tai Li, Trix van Mierlo, Nicolas Bencherki, Boukje Cnossen, Leonhard Dobusch, Richard Weiskopf, the anonymous reviewers, and especially special issue editors Stefania Milan and Davide Beraldo for their insightful feedback at various stages of this research.

## Disclosure statement

No potential conflict of interest was reported by the author(s).

## References

Ahrne, G., and Brunsson, N. (2011). Organization outside organizations: the significance of partial organization. Organization, 18(1), 83–104. https://doi.org/10.1177/1350508410376256

Banjo, S., Lung, N., Lee, A., & Dormido, H. (2019, August 22). *How Hong Kong's leaderless protest army gets things done*. Bloomberg. https://www.bloomberg.com/graphics/2019-hong-kong-airport-protests/

Barbosa, S., & Milan, S. (2019). Do not harm in private chat apps: Ethical issues for research on and with WhatsApp. *Westminster Papers in Communication & Culture, 14*(1), 49–65. https://doi.org/10.16997/wpcc.313

Bencherki, N., & Cooren, F. (2011). Having to be: The possessive constitution of organization. *Human Relations, 64*(12), 1579–1607. https://doi.org/10.1177/0018726711424227

Bennett, W. L., & Segerberg, A. (2012). The logic of connective action. *Information, Communication & Society, 15*(5), 739–768. https://doi.org/10.1080/1369118X.2012.670661

Benoit-Barné, C., & Cooren, F. (2009). The accomplishment of authority through presentification: How authority is distributed among and negotiated by organizational members. *Management Communication Quarterly, 23*(1), 5–31. https://doi.org/10.1177/0893318909335414

Beraldo, D. (2020). Movements as multiplicities and contentious branding: Lessons from the digital exploration of #Occupy and #Anonymous. *Information, Communication & Society, 25*(8), 1098–1114. https://doi.org/10.1080/1369118X.2020.1847164

Beraldo, D., & Milan, S. (2019). From data politics to the contentious politics of data. *Big Data & Society, 6*(2), 205395171988596. https://doi.org/10.1177/2053951719885967

Cammaerts, B. (2012). Protest logics and the mediation opportunity structure. *European Journal of Communication, 27*(2), 117–134. https://doi.org/10.1177/0267323112441007

Cooren, F., Bencherki, N., Chaput, M., & Vásquez, C. (2015). The communicative constitution of strategy-making: Exploring fleeting moments of strategy. In D. Golsorkhi, L. Rouleau, D. Seidl, & E. Vaara (Eds.), *Cambridge Handbook of strategy as practice* (2nd ed., pp. 365–388). Cambridge University Press.

Cooren, F., Kuhn, T., Cornelissen, J. P., & Clark, T. (2011). Communication, organizing and organization: An overview and introduction to the special issue. *Organization Studies, 32*(9), 1149–1170. https://doi.org/10.1177/0170840611410836

Dobusch, L., & Schoeneborn, D. (2015). Fluidity, identity, and organizationality: The communicative constitution of *Anonymous*: Fluidity, identity, and organizationality. *Journal of Management Studies, 52*(8), 1005–1035. https://doi.org/10.1111/joms.12139

Faraj, S., & Azad, B. (2012). The materiality of technology: An affordance perspective. In P. M. Leonardi, B. A. Nardi, & J. Kallinikos (Eds.), *Materiality and organizing: Social interaction in a technological world* (pp. 237–258). Oxford University Press.

Fayard, A.-L., & Weeks, J. (2014). Affordances for practice. *Information and Organization, 24*(4), 236–249. https://doi.org/10.1016/j.infoandorg.2014.10.001

Flesher Fominaya, C. (2010). Central concepts and debates: Collective identity in social movements. *Sociology Compass, 4*(6), 393–404. https://doi.org/10.1111/j.1751-9020.2010.00287.x

Gerbaudo, P., Marogna, F., & Alzetta, C. (2019). When "positive posting" attracts voters: User engagement and emotions in the 2017 UK election campaign on Facebook. *Social Media + Society, 5*(4), 205630511988169. https://doi.org/10.1177/2056305119881695

Gerbaudo, P., & Treré, E. (2015). In search of the 'we' of social media activism: Introduction to the special issue on social media and protest identities. *Information, Communication & Society, 18*(8), 865–871. https://doi.org/10.1080/1369118X.2015.1043319

Gibson, J. J. (1986). *The ecological approach to visual perception*. Lawrence Erlbaum.

Goffman, E. (1959). *The presentation of self in everyday life*. New York, NY: Anchor Books.

Gordon, L. (2002). Social movements, leadership, and democracy: Toward more utopian mistakes. *Journal of Women's History, 14*, 102–117. https://doi.org/10.1353/jowh.2002.0043

Granovetter, M. (1978). Threshold models of collective behavior. *American Journal of Sociology, 83*(6), 1420–1443. https://doi.org/10.1086/226707

Haug, C. (2013). Organizing spaces: Meeting arenas as a Social movement infrastructure between organization, network, and institution. *Organization Studies, 34*(5–6), 705–732. https://doi.org/10.1177/0170840613479232

Hensby, A. (2017). Open networks and secret Facebook groups: exploring cycle effects on activists' social media use in the 2010/11 UK student protests. *Social Movement Studies, 16*(4), 466–478. https://doi.org/10.1080/14742837.2016.1201421

Jasper, J. M. (1998). The emotions of protest: Affective and reactive emotions in and around social movements. *Sociological Forum*, *13*(3), 397–424. https://doi.org/10.1023/A:1022175308081

Kitchin, R., & Lauriault, T. P. (2015). Small data in the era of big data. *GeoJournal*, *80*(4), 463–475. https://doi.org/10.1007/s10708-014-9601-7

Kow, Y. M., Nardi, B., & Cheng, W. K. (2020). Be water: Technologies in the leaderless anti-ELAB movement in Hong Kong. *Proceedings of the 2020 CHI Conference on Human Factors in Computing Systems*, 1–12.

Lee, F. (2020). Solidarity in the anti-extradition bill movement in Hong Kong. *Critical Asian Studies*, *52*(1), 18–32. https://doi.org/10.1080/14672715.2020.1700629

Lee, F. L. F., & Chan J. M. (2016). Digital media activities and mode of participation in a protest campaign: a study of the Umbrella Movement. *Information, Communication & Society*, *19*(1), 4–22. https://doi.org/10.1080/1369118X.2015.1093530

Melucci, A. (1985). The symbolic challenge of contemporary movements. *Social Research*, *52*(4), 789–816.

Melucci, A. (1989). *Nomads of the present. Social movements and individual needs in contemporary society*. Hutchinson Radius.

Melucci, A. (1996). *Challenging codes: Collective action in the information age*. Cambridge University Press.

Milan, S. (2015a). From social movements to cloud protesting: The evolution of collective identity. *Information, Communication & Society*, *18*(8), 887–900. https://doi.org/10.1080/1369118X.2015.1043135

Milan, S. (2015b). When algorithms shape collective action: Social media and the dynamics of cloud protesting. *Social Media + Society*, *1*(2), 205630511562248. https://doi.org/10.1177/2056305115622481

Milan, S., & van der Velden, L. (2016). The Alternative Epistemologies of Data Activism. *Digital Culture & Society*, *2*(2), 57–74. https://doi.org/10.14361/dcs-2016-0205

Milan, S., & Barbosa, S. (2020). Enter the WhatsApper: Reinventing digital activism at the time of chat apps. *First Monday*, *25*(12). https://doi.org/10.5210/fm.v25i12.10414

Morrison, A. (2019, August 22). How Hong Kong's protests are affecting its economy. *The Conversation*. https://doi.org/10.14361/dcs-2016-0205

Polletta, F. (2006). *It was like a fever: Storytelling in protest and politics*. University of Chicago Press.

Polletta, F., & Jasper, J. M. (2001). Collective identity and social movements. *Annual Review of Sociology*, *27*(1), 283–305. https://doi.org/10.1146/annurev.soc.27.1.283

Poon, M., & Kohlberger, M. (2022). Twitter as a leadership actor—A communication as constitutive of organizing perspective on a 'leaderless' social movement. *Leadership*, *18*(5), 656–679. https://doi.org/10.1177/17427150221107271

Rogers, R. (2020). Deplatforming: Following extreme Internet celebrities to telegram and alternative social media. *European Journal of Communication*, *35*(3), 213–229. https://doi.org/10.1177/0267323120922066

Schoeneborn, D., Kuhn, T. R., & Kärreman, D. (2019). The communicative constitution of organization, organizing, and organizationality. *Organization Studies*, *40*(4), 475–496. https://doi.org/10.1177/0170840618782284

Taylor, J. R., & Cooren, F. (1997). What makes communication 'organizational'? *Journal of Pragmatics*, *27*(4), 409–438. https://doi.org/10.1016/S0378-2166(96)00044-6

Ting, T. (2020). From 'be water' to 'be fire': Nascent smart mob and networked protests in Hong Kong. *Social Movement Studies*, *19*(3), 362–368. https://doi.org/10.1080/14742837.2020.1727736

Treem, J. W., & Leonardi, P. M. (2012). Social media use in organizations: Exploring the affordances of visibility, editability, persistence, and association. *Annals of the International Communication Association*, *36*(1), 143–189. https://doi.org/10.1080/23808985.2013.11679130

Treré, E. (2012). Social movements as information ecologies: Exploring the coevolution of multiple internet technologies for activism. *International Journal of Communication*, *6*(19). https://ijoc.org/index.php/ijoc/article/view/1681/797

Treré, E. (2015). Reclaiming, proclaiming, and maintaining collective identity in the #YoSoy132 movement in Mexico: An examination of digital frontstage and backstage activism through social media and instant messaging platforms. *Information, Communication & Society, 18*(8), 901–915. https://doi.org/10.1080/1369118X.2015.1043744

Treré, E. (2020). The banality of WhatsApp: On the everyday politics of backstage activism in Mexico and Spain. *First Monday, 25*(1). https://doi.org/10.5210/fm.v25i12.10404

Vásquez, C., Bencherki, N., Cooren, F., & Sergi, V. (2018). From 'matters of concern' to 'matters of authority': Studying the performativity of strategy from a Communicative Constitution of Organization (CCO) approach. *Long Range Planning, 51*(3), 417–435. https://doi.org/10.1016/j.lrp.2017.01.001

Vincent, D. (2019, June 30). *How apps power Hong Kong's "leaderless" protests*. BBC. https://www.bbc.com/news/technology-48802125

Weiskopf, R. (2020). Algorithmic decision-making, spectrogenic profiling, and hyper-facticity in the age of post-truth. *Le Foucaldien, 6*(1), 1–37. https://doi.org/10.16995/lefou.62

Wertsch, J. V. (2012). Narrative tools and the construction of identity. In M. Schultz, S. Maguire, A. Langley, & H. Tsoukas (Eds.), *Constructing identity in and around organizations* (pp. 128–146). Oxford University Press.

Zammuto, R. F., Griffith, T. L., Majchrzak, A., Dougherty, D. J., & Faraj, S. (2007). Information technology and the changing fabric of organization. *Organization Science, 18*(5), 749–762. https://doi.org/10.1287/orsc.1070.0307

# PROFILE: *revisiting the social movement society* in a time of datafication

David S. Meyer

**ABSTRACT**
Revisiting the concept of the social movement society, we see the development of new media and datafication creating unanticipated challenges and constraints, in effect, new and different opportunities. The wired world reduces the costs of spreading information and mobilizing action, but also creates additional risks. I review the premises and blindspots of the 'social movement society' hypothesis, initially offered by David S. Meyer and Sidney Tarrow,' and then consider how new technologies affect strategies and possibilities for activists, as well as stakes and targets of activists. The conclusion raises issues of inequality and representation.

When Meyer and Tarrow (1998) proposed the idea of a 'social movement society,' first published in 1998, we ostentatiously suggested that that we were describing the politics of a new century. But predictions are a difficult game, particularly about the future, and we missed badly on a few essential issues, some of which are redressed here. It's not that we were wrong about *everything*, but rather that we failed to see the fundamental changes in the political arena wrought by technological advances. Perhaps even more significantly, we did not anticipate the reactions to the trends that we correctly identified. Importantly, new advances in technology change not only the means, but the stakes and the critical actors in politics, contentious and otherwise (Tilly & Tarrow, 2015). The following chapters lay out some of the changes and map strategies for making sense of the new world as it emerges.

One of the difficult delights of an academic life is getting to revisit the mistakes and misjudgments of the past, our own and those of others (e.g., Meyer & Pullum, 2015a, b). The movement society hypothesis animated a good deal of important research and a reexamination of contemporary political life as well as considered speculation about prospects for the future (see, e.g., Soule & Earl, 2005). And some of the challenges we identified in the last century remain. In this regard, the focused consideration on the consequences of datafication is not only welcome, but necessary. Here, I want to consider some of the implications of increased datafication on the emergence, trajectory of contention in contemporary politics. I begin by reviewing the basic elements of the social movement society hypothesis. After offering a brief overview of what I think we got right and what we missed, I point to a few critical issues that come from a level of datafication that we did not anticipate. I conclude with a call for more research like that represented in this issue.

## The social movement society

The fundamental premise of the social movement society hypothesis was that protest, often seen as the last resort of those who were otherwise excluded from meaningful participation in the political system, had become ubiquitous in advanced industrialized societies (Meyer & Tarrow, 1998). Earlier scholarship had seen protest as the recourse and resource of those who could not win otherwise (e.g., Lipsky, 1970). But protest was adopted as a tactic by both winners and losers in the political system, used both to augment their influence and to challenge their opponents. Rather than episodic, social movement organization and events were almost constant. Moreover, and clearly related, the organization of protest events and organizations that support such events had become largely routinized and professionalized. Experienced and paid professionals staffed so-called social movement organizations (e.g., Staggenborg, 1988), and tended to rely on a familiar – and limited – repertoire of action that was heavily weighted toward less confrontational and less disruptive actions, which authorities learned to manage. Professional organizers picked routines and strategies that were sustainable and modular – movable across issues. Greater numbers of people protested for a wide variety of causes, but the protests themselves would be less disruptive, threatening, and ambitiously comprehensive in their claims. Professionals came to their task with different concerns than citizen amateurs, who might be more focused on immediately grievances than organizational maintenance or stable relationships with authorities – and those membership organizations have, generally, atrophied (Skocpol, 2013).

Our vision of the future of protest politics came with several cautions. First, the proliferation of protest campaigns on multiple sides of every issue would make governance more difficult. Governments could easily face substantial protest movements no matter what policies they adopted, and supporters of those movements in parliaments would be emboldened in opposition – knowing that they could likely benefit from support in the streets. Governmental paralysis came with another consequence: less advantaged constituencies in the political system – exactly those thought to benefit from social protest movements – would be neglected by governments and – absent extensive resources, substantial organizations, and professional advocates – unable to mobilize on their behalf. All of this would lead to growing inequality.

I'd say the basic premises of the movement society hypothesis were right, but it missed the emergence of populist movements on the left and the right over the past 20 years, movements that have rocked politics and governance in substantial ways that will play out over a long period of time. From a contemporary vantage point, such movements were possible because of changes in communication technologies, including social media, which were already in process when we were writing. Those changes have shaken up governance and politics in substantial ways, dramatically altering the contours of any social movement society.

## It's a wired world

The prediction about the professionalization of politics was premised on the idea that the tasks of disseminating information and linking and mobilizing people were resource intensive, and required both technical expertise and institutional infrastructure. We saw the continuation of a classic model of political mobilizing, in which professional organizers took a message to existing communities and invited them to engage in focused political activity which often included protest. But the world changed. The creation of almost global communication networks carrying messages laterally changed the dynamics of organizing. Entrepreneurial organizers could spread a message on a variety of social media platforms below the level of any media gatekeepers, providing messages about politics, policy, and organizing that were unmediated, unedited, and unchecked, allowing information and misinformation to spread across a variety of networks that could be larger and more insulated than in the past (e.g., see Bennett & Segerberg, 2013; Beraldo & Milan, 2019; Earl & Kimport, 2011; Tufekci, 2017).

But this is but one slice into the way in which the growth of media has fundamentally changed the stakes and the strategies of collective action in the contemporary world. Let me suggest a few ways in which the new media universe has fundamentally altered the political arena, starting with the strategies and possibilities for organizing that are now available, and then considering how the stakes and grievances of grassroots organizers have also changed. Every advantage comes with a caution, and much work is needed to make sense of the longer-term implications of the new politics.

### *Strategies and possibilities*

The growth of the Internet and the proliferation of extensive social media platforms adds more than a new wrinkle to available means for organizing dissent. *The Social Movement Society* imagined conventional professional organizers who would have resources to advertise, employ canvassers, and maintain offices with telephones and fax machines used to coordinate ongoing campaigns. At this point, however, while such professionalism remains a feature of the organizing world, it is by no means the exclusive mode of organizing. Online organizers can reach and often mobilize audiences without much professional expertise or substantial resource bases.

The Occupy campaign provides a useful illustration. Suggested by an activist magazine, local organizers in New York City planned an event in the financial capital of the United States, producing a pretty poster and a vague but ambitious strategy. The initial turnout was small, and generated a tiny bit of attention in mainstream media. But in setting up Occupy Wall Street, a new media team, armed primarily with laptops, provided their own press that largely bypassed mainstream media to reach a national audience. Setting up several livestreams, Occupy press activists were able to circulate not only commentary and invective, but videos of meetings, protests, and engagement with locals, including merchants and police. The messages were picked up around the country and, indeed, around the world, where other groups of activists were able to pick up on both the grievances and the strategies. To some extent, resources followed the activism, but surely the most important resource was engaged activist support, which produced hundreds of occupations of all sizes and slightly different foci across the United States and around the world.

Occupy was unusual in its diffusion, but it was hardly the first group to create a kind of left-wing populism linked to a dramatic political strategy. Activists cited the Arab Spring nearly a year earlier as inspiration, and they had their own pictures and stories to make the case. They could have also acknowledged broad occupation campaigns in Spain, Israel, and in Madison, Wisconsin, earlier in the year with similar concerns. The point is that new technologies allowed the spread of a set of concerns and a tactic (occupation of public spaces) broadly, without much in the way of professional engagement or financial support. Skillful use of social media to project causes and events allowed distant supporters to engage the issue, emulate tactics, and provide support. In an odd example, thousands of supporters across the country – indeed, around the world – ordered pizzas from Madison restaurants to be delivered to the protesters occupying the state capitol (Shiner, 2011). On the other hand, the collapse of the Occupy organizations in the wake of a coordinated campaign to end the occupations underscores the weaknesses inherent in the amateur approach.

Occupy was unique in scale, but the use of the Internet to disseminate information and strategies to encourage and steer organizers is now widespread. Importantly, grassroots media, deployed on established social media platforms, allows news of outrages (like police killings) and actions to spread without the cooperation of mainstream legacy media in newsprint or on television. A hashtag or online petition can provide visibility to an issue, a bit of political education, and a display of unity all attached to a set of concerns. The relatively spare resource commitment needed to seed and stoke movements has allowed left and right populisms to flourish.

Both left and right groups were able to use social media to organize local campaigns but, importantly, the constituencies activated are no longer bounded geographically. The seemingly deep connections forged online extend past neighbors and local communities, creating communities that are unified around a set of beliefs or symbols, and that are largely unchecked by internal dissent. Exit, in Hirschman's (1970) terms, is relatively easy and frictionless, as is finding an alternative online group and an identity more aligned with an imagined ideal.

Moreover, the capacity to spread unmediated, unsubstantiated, and often unprincipled assertions about science, politics, or anything else, produces a kind of corruption of information, deepening commitment within a movement, while simultaneously creating an insularity preventing democratic dialogue inside and outside that movement (Polletta 2002). A social movement creates its own system of information and ideological support, a kind of epistemic ecosystem. In online communications, it's easier to posture than to discuss, leading to, I suspect, geographic dispersion combined with ideological homogeneity within a community.

Bypassing institutional filters for information allows dissemination of damaging information about political opponents, including states. Legacy media, like newspapers (even if now online) and broadcast networks, adhere more or less to journalistic norms that don't constrain activists. When Defense analyst Daniel Ellsberg took the Pentagon Papers, a secret history of the Vietnam War, to mainstream papers, editors and lawyers pored over the texts, and argued about just which details to withhold while serving the public interest. In contrast, Wikileaks and its founder Julian Assange, vigorously disavowed such a responsibility, promising to publish everything they got online, leaving it to others to sort through and decide how to use. This approach creates cascades of information, of varying importance or potential damage. It likely will still the accumulation of public records altogether.

Ubiquitous information recording (e.g., cell phone cameras) and dissemination (e.g., livestreams) tools afford activists the prospect of citizen journalism in a way not previously imagined. The explosion of Black Lives Matter protests against police violence in the United States in 2020 was possible because bystanders were able to record a particularly outrageous instance of police violence, the murder of a Black man in custody outside a convenience store (Caren et al., 2023). It's not that such instances were new in 2020, but that the visceral impact of a video witnessed on a phone or home computer far outstripped even a graphic description – which could be disputed by authorities. It's somewhat easier to expose and to publicize egregious cases of injustice.

It's not only grassroots populist activists who can access social media to spread a message and to organize. Powerful actors inside and outside government also enjoy the same access and can deploy their messages to larger audiences at the outset (Tilly, 2008). They too can bypass conventional media and well-established standards of journalism to promote preferred messages about themselves, their opponents, and matters of policy. There is no one or set of players with a monopoly on new communications technology, but states and large businesses enjoy at least as much power and influence in new media as grassroots activists – actually, much more (Tarrow, 2022). The new communications and organizational capacity provides powerful resources for new populists and pseudo-populists. America's Donald Trump, Brazil's Jair Bolsonaro, and the UK's Boris Johnson provide handy examples of right wing populists who've proved adept at using social media to bypass both conventional political structures and mainstream media. In addition to spreading misinformation, the proliferation of bad or unreliable reporting leads to a kind of Gresham's law for information, undermining confidence in all journalistic enterprises. Ultimately, declining respect for reliable and intersubjective information is bad for democracy, and indeed, for effective governance altogether.

Command of social media is also a sharp threat to political opponents. Activists and institutional actors can target political enemies, even individuals not normally enmeshed in public debates. In addition to disseminating misinformation, purposeful actors can disclose accurate information that is normally private, 'doxing' opponents while offering a general encouragement to unidentified allies that a home address or phone number is something that can be used. A prosecuting attorney whose home address is disclosed can face harassment and threats for worse. Minimally, targeted opponents can be ridiculed and harassed online, never knowing exactly which hostile threats or abusive language might be a precursor to a physical assault. Public health officials in the United States who advocated unpopular restrictions on economic and social life in the service of fighting Covid faced unusual political pressures that included online threats to their safety and that of their families, as well as disruptive protests outside their homes. They were unable to count on the support of elected officials who faced similar pressures; scores of officials resigned, retired, or were fired, and finding qualified replacements was difficult (Associated Press, 2020).

The new wired world also affords the possibility of attacks on others' communication capacity. Long ago, activists learned how to stage denial of service attacks on wired businesses and services, which could paralyze a company, state agency, or religious or political group for a time (e.g., George-Cosh, 2008). More recently, activists and others have learned to use ransomware to harass or blackmail their targets. Motivations are often purely commercial – attackers look for money – but there can be political purposes as well. It's quite likely that activists, pirates, and corporate actors will continue to leapfrog each other in developing attacks and defenses.

### *Stakes and targets*

The growth of media and the accumulation and sorting of extensive information on individuals also fundamentally alters the stakes of social justice organizing. An increasingly wired world, one in which closed circuit television cameras are ubiquitous and tracking individual movements via drones or, better, phones, is one in which anonymity becomes increasingly difficult. Surveillance of even the mildly suspect has become easier, almost routine, and whatever tools insurgents can access are also available, often with greater consequence and capacity, to both states and private actors like corporations. Social media companies, which make money by selling advertisements, routinely track – with varying gestures to anonymization—the connections, posts, and searches of their users, making it easier to sell target populations to advertisers. Tracking the location of cell phones is a service available to customers, and certainly to states and corporations as well. Enhanced surveillance puts activists at greater peril when they engage in – or even plan to stage – events that challenge or disrupt the plans of states or companies. Of course, authorities can track and prosecute on the basis of beliefs as well as tactics, threatening to still selected elements of an increasingly distorted public space.

Surveillance and tracking of individuals' activities and consumption choices – online and off – also creates a valued commodity of lists: lists of people who might be interested in a product, a political candidate, or a cause; lists of people who might be engaging in criminal enterprises or in political causes (Zuboff, 2019). The commodification of individual lives through commercial lists presents not only an obstacle to activists, but also a cause – privacy—to try to rally around. Activists share strategies for 'going dark,' that is, limiting their exposure to political or corporate collectors of data, by more and less sophisticated means that begin with leaving a cell phone at home. Privacy invaders include not only states, but also companies, and the lines of defense against those invasions are staffed by states and supranational institutions which face their own difficulties in protecting themselves and their citizens. The dangers are clearly presented in ongoing struggles about democracy in Hong Kong, where China's tools of surveillance have been instituted, and activists are appropriately fearful of increasingly harsh prosecutions and punishments (Human Rights Watch, 2021). At the same time, they depend upon publicizing their plight and their cause outside of Hong Kong, hoping to find some protection with visibility, and to mobilize outside supporters (see, e.g., Keck & Sikkink, 1998)

A new set of political cleavages based on the availability of data has emerged. While some activists work to limit the collection of commodified data and surveillance, others work to promote the availability of information to promote informed democratic engagement. This includes citizen journalism, publishing state secrets and offensive conduct, but also breaking down the paywalls and subscription protocols that limit access to commodified information. Take, for example, the case of Aaron Swartz, a programmer and hacktivist whose key cause was democratizing data (Scheiber, 2013). Swartz effectively *liberated* court records and, later, academic journal articles, that authorities and businesses typically charged readers to access. Swartz committed suicide when facing trial for violating JSTOR's (a digital archive of academic research) privacy rights by disseminating access to hundreds of thousands of journal articles. He became a martyr to the cause of democratizing information, one whose story has been retold in all kinds of dramatic ways, including a documentary film, *The Internet's Own Boy: The Story of Aaron Swartz* (2014). Although the prosecution and exceptionally harsh punishment Swartz faced (possible fines exceeding $1 million, prison sentence of up to 50 years) may well have deterred open access hacktivists in the United States, the academic paywall has essentially been busted by a follow-on effort, Sci-Hub (https://sci-hub.se/), which provides free access to millions of academic papers. Founded in Kazakhstan and hosted somewhere distant from enforceable copyright law, the resource demonstrates the prospects and difficulties of the cause of democratizing data. Sci-Hub promotes its effort at digital privacy as part of a fundamentally democratic mission, announcing on its website, 'We fight inequality in knowledge access across the world. The scientific knowledge should be available for every person regardless of their income, social status, geographical location and etc. Our mission is to remove any barrier which impeding the widest possible distribution of knowledge in human society!' (http://www1.chim.unifi.it/u/signo/sci-hub.html)

The battles over datafication revolve around making some information more accessible to a wider variety of actors, while also preventing the collection of other sorts of data – primarily, about persons.

## Conclusion/Introduction

Datafication means the creation of new grievances and causes, new tools for activism, and novel coalitions of activists and entrepreneurs who may work together on various projects. How this new world works to shape potential activist and citizen identities and strategies for engagement is very much an open question for both scholars and citizens. Sid Tarrow and I certainly didn't see these possibilities when we published *The Social Movement Society*. The chapters that follow engage these critical issues and more.

The Social Movement Society was premised on a vision of saturating one form of social and political organization. Social media, or 'digital society' has changed all of that, creating a great deal of largely unmediated space, while simultaneously reducing the centrality and importance of the older spaces – which mostly still exist. In addition to the old universe of interest groups, mobilizing political parties, and emergent social movements based on face-to-face contacts remains. But there's more: an additional dimension is a host of targets and grievances centered in corporate actors, plus a flatter, more horizontal world in which individuals self-educate, self-select, and self-mobilize with unmediated sources of information. In some ways, this reality suggests a cautionary warning like the old one Kornhauser (1959) offered about the absence of intermediary associations leading to the creation of 'mass society' and the emergence of fundamentally undemocratic social movements. Kornhauser famously failed to anticipate the emergent social movements of the 1960s organized around student activism, an anti-war movement, and civil rights in particular – all movements that were built on pre-existing networks and all of which developed sustainable organizations. Kornhauser's caution may be more applicable to the current moment. There's a great deal of unfiltered 'information,' unmediated horizontal ties that are exploitable, reduced costs for organizing, and reduced authority of the state – and all sorts of other institutions – to adjudicate disputes.

There's much work to be done, by both activists and analysts, to make sense of this new world, and to find ways to create structures that maintain some sort of effective democratic governance. The essays collected here start us on that task.

## Disclosure statement

No potential conflict of interest was reported by the author(s).

## ORCID

David S. Meyer http://orcid.org/0000-0001-8511-9307

## References

Associated Press. (2020). At least 181 public health leaders nationwide have resigned, retired or been fired amid pandemic. https://ktla.com/news/coronavirus/at-least-181-public-health-leaders-nationwide-have-resigned-retired-or-been-fired-amid-pandemic/

Bennett, W. L., & Segerberg, A. (2013). *The logic of Connective action digital media and the personalization of Contentious politics*. CUP.

Beraldo, D., & Milan, S. (2019, July–December). From data politics to the Contentious politics of data. *Big Data & Society*, *6*(2), 1–11. https://doi.org/10.1177/2053951719885967

Caren, N., Andrews, K. T., & Nelson, M. H. (2023). Black lives matter protests and the 2020 presidential election. *Social Movement Studies*, 1–18. https://doi.org/10.1080/14742837.2023.2216652Caren

Earl, J., & Kimport, K. (2011). *Digitally enabled social change: Activism in the Internet age*. MIT Press.

George-Cosh, D. (2008, January 26). Online group declares war on scientology. *National Post*. https://archive.ph/20080128145858/http://www.nationalpost.com/news/canada/story.html

Hirschman, A. O. (1970). *Exit, Voice, loyalty*. Harvard University Press.

Human Rights Watch. (2021, June 25). Dismantling a free society: Hong Kong one year after the national security law. Report. https://www.hrw.org/feature/2021/06/25/dismantling-free-society/hong-kong-one-year-after-national-security-law

Keck, M. E., & Sikkink, K. (1998). *Activists beyond borders: Advocacy networks in international politics*. Cornell University Press.

Kornhauser, W. (1959). *The politics of mass society*. Free Press.

Lipsky, M. (1970). *Protest in city politics*. Rand-McNally.

Meyer, D. S., & Pullum, A. (2015a). Reconsidering the social movement society in the new century. In H. Ramos & K. Rogers (Eds.), *Protest and politics: The promise of social movement societies* (pp. 21–43). University of British Columbia Press.

Meyer, D. S., & Pullum, A. (2015b). The social movement society, the tea party, and democratic deficits. In Lee, C. W. M. McQuarrie & T. W. Edward (Eds.), *Democratizing inequalities: Pitfalls and unrealized promises of the new public participation* (pp. 204–221). New York University Press.

Meyer, D. S., & Tarrow, S. (Eds). (1998). *The social movement society: Contentious politics for a new century*. Rowman & Littlefield.

Polletta, F. (2002). *Freedom is an endless meeting: Democracy in American social movements*. University of Chicago Press.

Scheiber, N. (2013, February 13). The inside story of why Aaron Swartz broke into MIT and JSTOR. *The New Republic*. https://newrepublic.com/article/112418/aaron-swartz-suicide-why-he-broke-jstor-and-mit

Shiner, M. (2011, February 20). From Cairo to Madison, some pizza. *Politico*. https://www.politico.com/story/2011/02/from-cairo-to-madison-some-pizza-049888

Skocpol, T. (2013). *Diminished democracy: From membership to management in American civic life*. University of Oklahoma Press.

Soule, S., & Earl, J. (2005). A movement society evaluated: Collective protest in the United States, 1960-1986. *Mobilization: An International Quarterly*, *10*(3), 345–364. https://doi.org/10.17813/maiq.10.3.730350353753l022

Staggenborg, S. (1988). The consequences of professionalization and formalization in the pro-choice movement. *American Sociological Review*, *53*(4), 585–605. https://doi.org/10.2307/2095851

Tarrow, S. (2022). *Power in movement: Social movements and Contentious politics* (4th ed.). CUP.

Tilly, C. (2008). *Contentious performances*. Cambridge University Press.

Tilly, C., & Tarrow, S. (2015). *Contentious politics* (2nd ed.). Oxford University Press.

Tufekci, Z. (2017). *Twitter and Tear Gas: The power and fragility of networked protest*. Yale University Press.

Zuboff, S. (2019). *The age of surveillance capitalism*. Profile Books.

# Data as narrative: contesting the right to the word

Nick Couldry

**ABSTRACT**
This guest contribution to the Contentious Data special issue asks: What are the fundamental processes underlying the possibility of data activism? It argues that, if like everything else, social movements are being datafied, this operates on at least four levels: a change in the general conditions under which all social movements operate; data becoming either the specific or general object of activism; and finally, data becoming crucial to practices of movement resistance. Underlying this is a further pattern, that data as a narrative are increasingly an important aspect of contestation in contemporary politics. I interpret this general phenomenon through the lens of the social theory of Alberto Melucci and the leader of the Zapatistas movement, Subcomandante Marcos.

> Once it is recognized that the power of information is essentially the power of naming, we can set out on the enormous task of redefining the "right to the word" that is called for in the information age (Melucci, 1996, p. 228)

In recent years, a new word has entered our research vocabulary: (in English) datafication (Van Dijck, 2014). Datafication gestures towards something enormous: a general transformation of the social fabric, of the very stuff from which social contexts, social interfaces, social interactions and social space more generally are made. The transformation happens, quite simply, because the elements, inputs and scenes of social life are themselves becoming data. Many areas of social and economic production have been transformed (to name a few: agriculture, health, education, the workplace, logistics, government). It is no surprise that social movements are being transformed too. This applies, whatever macro-theory one adopts to make sense of the advance of datafication, for example, surveillance capitalism (Zuboff, 2019) or data colonialism (Couldry & Mejias, 2019).

To say, therefore, that social movements are datafied is merely a truism, as Milan and Berardo (2024) note. That truism tells us nothing, unless we clarify in what specific sense social movements are *distinctively* being datafied, compared with other areas of daily life. In this short article, I will try to provide a brief overview of some of the ways in which this is happening.

### *Four levels of social movements' datafication*

We can identify at least four levels on which social movements as distributed human groups that work for transformative social change are being transformed by datafication. The articles in this Contentious Data special issue illustrate each of them (Milan et al., 2024).

*First*, datafication transforms the general *conditions* under which any movement, whether it is calling for social change or not, can operate. The circumstances of right-wing populist movements that work to restore an imagined exclusivist utopia and so reverse decades of hard-fought progress on human rights and cultural/social recognition, as in today's Argentina, Brazil, Germany, Poland, and US are being transformed by datafication just as much as are those of the movements that seek to resist those populists, and indeed as much as the conditions of groups not concerned with any form of social change (local choirs, sports clubs, yoga schools, and so on).

One general aspect of this effect is what has been called 'institutionalization' (Couldry & Hepp, 2016, pp. 217–18): that is, the growing pressure to build data collection and data use into mechanisms of mobilisation and group formation, that, in a broader sense (given the complex technological infrastructure required), involves the institutionalization of social movements. One can see datafied forms of institutionalization as a constraint on action (new types of data skills are needed now by movements, which may cost money), but also as an opportunity. Poon (2024) argues for the latter in her article about how activists in Hong Kong contest the Chinese state by conducting important organizational work via an encrypted network such as Telegram. Such opportunity only arises, however, within a wider context where the surveillance risks for social movements, both today and in the future (that is, as their archived communications on their phones and elsewhere become vulnerable to surveillance by the state), are on average rising. The balance of risks from the massive increase in the surveillance and encryption capacities of both states and corporate communication platforms will however play out differently in different locations, depending on the states and platforms relevant to a particular social movement.

A different way in which the general conditions of social movements are changed by datafication was discussed by Wright (2022). As the sorts of things that social movements target for contestatory action become not social policies or broader social or economic injustices, but increasingly the information infrastructures in which all social action is today entangled, the frequency of political opportunities for intervention becomes rarer. The reasons relate not so much to the detailed features of data, as to the dislocation between the temporality of infrastructure (such as the infrastructure of the internet backbone) and the temporality of mainstream political struggle: this dislocation is huge and, for some types of deep data activism, absolute. The result is to change the rhythm of social movement activism, or at least those versions of it that have to challenge information infrastructures of various sorts.

The next two levels on which datafication transforms social movements are more specific. They relate to the distinctive *objects* of social movement action. A lot depends here on the angle of vision of particular social movements, which of course varies enormously and for reasons linked to the particular societies in which they are embedded.

One way in which data is transforming social movements is by shifting the *specific object* which they contest. Milan and Berardo (2024) give some clear examples of this: battles for data justice in the US over how specific groups and individuals are represented in powerful state and corporate databases, battles over who controls the operation and application of those databases, and so on. Alternatively, cases where the workings of data gathering are contested, as with the Indian government's Aadhaar identity system. Muravyov's article (Muravyov, 2022) gives a particular interesting example of how data on the Russian transport system is a direct site of activist contention, and serves as an entry-point to wider contestation of the information politics of Russia's authoritarian state. Sometimes protesting about data means challenging the means of its extraction and processing and the larger narratives generated from it; sometimes, as Li and Whitworth (2023) bring out, it is a matter of reversing processes of data formation so as to turn the tables on a powerful actor, such as the Chinese state, receiving 'a dose of its own medicine' by becoming subject to sousveillance by Hong Kong activists. Note that data activism of this sort will sometimes be about algorithms (the automated calculative mechanisms for generating new data from existing data), but often it will operate at other levels: challenging the design of databases, the formulation of data categories and data aggregates, the linkage of otherwise unconnected databases.

A related, third, level where datafication transforms the daily circumstances of social movements is by reconstituting the *general object* of their activism. This happens when a movement is contesting not specific data, but the whole infrastructure (not necessarily technological) from which specific data are generated. This might be a state infrastructure (as in Russia or Hong Kong) or it might be a corporate infrastructure as with Google's attempts to build a fully datafied public space in Toronto, discussed by Charitsis and Laamanen (2022). The latter article shows well how, when datafication becomes a general target of movement action, important overlaps with wider social movements (movements that do not take data, whether generally or specifically, as their object) come into view. As one Canadian activist cited by Charitsis and Laamanen (2022) put it: 'we strongly believe that Toronto can win against Alphabet/Sidewalk Labs and their plan to take over our lives and spaces'.

The fourth level on which datafication changes social movements is perhaps the most discussed so far in the data activism literature. This is where data becomes intrinsic to social movements' methods of resistance, whatever the social process those movements are resisting. In a datafied society, data are a tool of social action in every aspect of daily life, from using a phone to setting up a group for collaboration, which means data are a site of agency by social movements. It is sometimes suggested that macro-theories of how we interpret datafication rule out attention to such agency (Lash, 2007), but this doesn't follow. Take the framework of data colonialism for example. It doesn't follow from this theory that there is no room for individual agency to resist it. On the contrary, the goal of developing such a theory can be seen as stimulating such agency and resistance, understanding data colonialism as a social order which is built *through* human agency on all scales, in which everyone is complicit, until, that is, they are not and they start resisting it (Couldry & Mejias, 2019, chapters 1 and 6). Data, in any case, are at root symbols, and, like all symbols, they are at some level contestable, that is, subject to reinterpretation: data categories can be used and turned against themselves, or at least against the intentions of their designers. (From the point of view of recognising agency, Zuboff's top-down model of surveillance capitalism as a 'coup from above' (Zuboff, 2015, p. 86), which operates without any consent or complicity may be a little different).

Datafication, therefore, is quite consistent with possibilities of agency. This is what Treré and Bonini (2022) set out to explore. Data can be involved as part of the basic institutionalization of movements, as we saw earlier, or as a specific tool and mechanism of resistance, for example where algorithms, and their modalities of automation, are contested or manipulated against their designers' intentions. Spain's 15 M movement which emerged in the aftermath of the 2007–2008 financial meltdown provides, as they say, a well-known example of this (see also Postill, 2014). One related point is however worth making: that what matters with agency in relation to datafication is the contestation of *data*, rather than necessarily algorithms. Algorithms are only one dimension of datafication, which perhaps raises a question about the usefulness of terms such as 'algorithmic activism'. Very often it is something about data other than the workings of the algorithm that is being contested (for example, the construction of data categories, their combination into larger data aggregates, the linking of previously unconnected databases), but such activism is no less important for that.

## *An underlying pattern: data as narrative*

Is there an underlying pattern beneath these four levels of how datafication transforms social movements that, if we can clarify it, might illuminate the field of datafied social struggle in a broader way? There is, I suggest, and that is the idea of data as narrative.

The possibility of data as narrative is implicit in the quote from the great theorist of 'new' social movements, Alberto Melucci, from which I started. Melucci made this point in the course of a more fundamental investigation of 'the field of conflict' (Melucci, 1996, p. 4) in contemporary social and political struggles. That field is profoundly transformed, Melucci argued with remarkable foresight, in a world where 'the real domination is today the exclusion from the power of naming' (Melucci, 1996, p. 182).

Melucci could not of course have anticipated today's world of social media, apps, and point-to-point surveillance and sousveillance, let alone the datafication of every aspect of daily life. But he already grasped the direction of travel towards a world where contestation of information and symbols would become central and transformative. As he put it already in his earlier book, *Nomads of the Present*, in late modern societies 'conflicts develop in those areas of the [social] system which are crucial for the production of information and symbolic resources, and which are subject at the same time to the greatest pressure to conform' (Melucci, 1989, p. 55). It is hard to imagine a better general formulation of the shift of social contestation and resistance over the past two decades via the uses of social media as an organising and mobilising tool, for example, in the misnamed 'Facebook revolutions' which occurred in Egypt and Tunisia in 2011 (Castells, 2012). For sure, Facebook is a symbolic resource, but also a site of huge pressure to conform, a site, if you like, where social order is reproduced.

For a more poetic and polemical expression of a similar point, we can turn to the leader of the Zapatista movement, a radical movement that in Mexico, from

1994 onwards, contested the long history of unequal land rights and exploitation of indigenous populations (Hayden, 2002). Its leader, Subcomandante Marcos, wrote:

> Power uses the word to impose his empire of silence. We use the word to renew ourselves. Power uses silence to hide his crimes. We use silence to listen to one another, to touch one another, to know one another. (Marcos, 2001, p. 84)

Marcos pits the Zapatistas' strategic use of the word and silence against their strategic use by large-scale power.

How do these broader visions of contemporary social movements fit with the analysis of four levels just developed? The first level was explicitly about how datafication changes the background conditions of social movements: this fits therefore with Melucci's general analysis, but does not have much to tell us about *how* datafied social movements contest the social. The next three levels however fit within Melucci's vision of the growing prominence of information and symbolic contestation in contemporary conflict. More specifically, the three levels of data-targeted, or data-enabled, social movements (compare Milan and Berardo's (2024) distinction between data as content and data as infrastructure, and Treré and Bonini's (2022) distinction between algorithmically-enabled and algorithmically-oriented activism) can, all of them, be seen as examples of the growing centrality of battles over 'the right to the word' in Melucci's phrase.

When we step back from the complex stack of automated information technologies that make contemporary data extraction possible, what *are* data but an attempt by corporations, or states or combinations of them, to read and speak to the world on their own terms and activists, in turn, to contest those privileged readings, that privileged speech? While activism is today, trivially, datafied, data-related practices of activism are attempts to contest those practices' representation of the world on some scale or other.

The core issue then, underlying all the complexities of term about data-related activism, is narrative. This remains true, even though a crucial aspect of all data is not just words, but quantification: the counting of myriad categories, formulated in computer code. Data must be contested, because it enforces *one* narrative of the world, in which are encoded particular interests and power relations. Activists, by contesting data, are contesting narratives of this sort. This is the specificity of data-related activism.

Framing the topics of this special issue (2024) in this broader way, however, brings out a further point: that it is important not to emphasise the technological details of what activists are now doing if that means losing sight of those broader issues of narrative. It is quite possible for activists to become too focussed on the micro-details of data strategy and fail to develop a wider narrative which effectively contests the whole state of society that needs to be challenged. What matters most, in the end, are not questions of data, but the articulations between data-related practices and the wider terrain of the social.

Put another way, what matters most is how, through data, the right to the word is today being fought for in new and important ways. Melucci's and Marcos' insights endure in ways they could not have expected. Let me end by recalling one of the deepest forms of data activism at work today: the contestation by Black activists, particularly in the US, of their exclusion from the basic practices and skill sharing of computing for much of computing's history. In response, as Charlton McIlwain has eloquently shown, they build a counter-practice of 'black software'. 'black software', he writes (McIlwain, 2020, p. 7), 'refers to the programs we desire and design computers to run. It refers to who designs the program, for what purposes, and what or who becomes its object or data'. In this way, data activism becomes integral to the very constitution of knowledge and politics today.

## Disclosure statement

No potential conflict of interest was reported by the author(s).

## References

Castells, M. (2012). *Networks of outrage and hope*. Polity.

Charitsis, V., & Laamanen, M. (2022). When digital capitalism takes (on) the neighbourhood: Data activism meets place-based collective action. *Social Movement Studies*, 1–18. https://doi.org/10.1080/14742837.2022.2123314

Couldry, N., & Hepp, A. (2016). *The mediated construction of reality*. Polity.

Couldry, N., & Mejias, U. (2019). *The costs of connection: How data colonizes human life and appropriates it for capitalism*. Stanford University Press.

Hayden, T. (Ed.). (2002). *The Zapatista reader*. Bold Type Books.

Lash, S. (2007). Power after hegemony: Cultural studies in mutation. *Theory, Culture & Society*, *24* (3), 55–78. https://doi.org/10.1177/0263276407075956

Li, Y.-T., & Whitworth, K. (2023). Coordinating and doxing data: Hong Kong protesters' and government supporters' data strategies in the age of datafication. *Social Movement Studies*, 1–18. https://doi.org/10.1080/14742837.2023.2178404

Marcos, S. (2001). *Our word is our weapon*. Serpent's Tail.

McIlwain, C. (2020). *Black software: The internet and racial justice, from the AfroNet to black lives matters*. Oxford University Press.

Melucci, A. (1989). *Nomads of the present: Social movements and individual needs in contemporary society*. Hutchinson Radius.

Melucci, A. (1996). *Challenging codes*. Cambridge University Press.

Milan, S., Beraldo, D., & Fominaya, C. F. (Eds.). (2024). Contentious data. The shifting space of social movement research in datafied societies. *Special issue*. Social Movement Studies.

Milan, S., & Berardo, D. (2024). Data in movement. The social movement society in the age of datafication. *Social Movement Studies*.

Muravyov, D. (2022). Doubt to be certain: Epistemological ambiguity of data in the case of grassroots mapping of traffic accidents in Russia. *Social Movement Studies*, 1–17. https://doi.org/10.1080/14742837.2022.2128327

Poon, M. (2024). Achieving organizationality through authorship affordances — a communicative episode of telegram polling from 2019 Hong Kong. *Social Movement Studies*, 1–22. https://doi.org/10.1080/14742837.2023.2296868

Postill, J. (2014). Democracy in an age of viral reality: A media epidemiography of Spain's indignados movement. *Ethnography*, *15*(1), 51–69. https://doi.org/10.1177/1466138113502513

Treré, E., & Bonini, T. (2022). Amplification, evasion, hijacking: Algorithms as repertoire for social movements and the struggle for visibility. *Social Movement Studies*, 1–17. https://doi.org/10.1080/14742837.2022.2143345

Van Dijck, J. (2014). Datafication, Dataism and Dataveillance: Big data between scientific paradigm and ideology. *Surveillance & Society*, *12*(2), 197–208. https://doi.org/10.24908/ss.v12i2.4776

Wright, J. (2022). "The future of the internet hangs in the balance": The perception and framing of political opportunity and threat in the contentious politics of data. *Social Movement Studies*, 1–18. https://doi.org/10.1080/14742837.2022.2123313

Zuboff, S. (2015). Big other: Surveillance capitalism and the prospects of an information civilisation. *Journal of Information Technology*, *30*(1), 75–89. https://doi.org/10.1057/jit.2015.5

Zuboff, S. (2019). *The age of surveillance capitalism*. Profile Books.

# Index

Note: Figures are indicated by *italics*. Tables are indicated by **bold**. Endnotes are indicated by the page number followed by 'n' and the endnote number e.g., 20n1 refers to endnote 1 on page 20.

Kolibri 360 GmbH
Ferdinand-Jühlke-Straße 7
99095 Erfurt, Deutschland
produktsicherheit@kolibri360.de